34th Eg-MRS International Conference

IOP Conference Series: Materials Science and Engineering
Volume 956

Cairo, Egypt
29 - 30 August 2020

Editor:

Arafa H. Aly

ISBN: 978-1-7138-3170-9
ISSN: 1757-8981

Printed from e-media with permission by:

Curran Associates, Inc.
57 Morehouse Lane
Red Hook, NY 12571

Some format issues inherent in the e-media version may also appear in this print version.

Institute of Physics
Dirac House, Temple Back
Bristol BS1 6BE UK

Phone: 44 1 17 929 7481
Fax: 44 1 17 920 0979

techtracking@iop.org

Additional copies of this publication are available from:

Curran Associates, Inc.
57 Morehouse Lane
Red Hook, NY 12571 USA
Phone: 845-758-0400
Fax: 845-758-2633
Email: curran@proceedings.com
Web: www.proceedings.com

TABLE OF CONTENTS

IOP Conf. Series: Materials Science and Engineering **956** (2020) 011001 doi:10.1088/1757-899X/956/1/011001

Proceeding of XXXIV International Conference of Material Sciences and Its Applications (XXXIV Eg-MRS) 2020

Preface

It is with great pleasure that I introduce the proceedings of the XXXV Eg-MRS International Conference, 29-30 August 2020, Cairo, Egypt. This conference was dedicated to technical issue related to material sciences research and its applications.

The objectives of the conference were not only to provide opportunities to the international scientists to present state of the art research in material sciences but also to provide a forum to discuss ideas for future directions in the field and explore possibilities for collaborative research projects.

The conference program included keynote, oral, and poster presentations from scholars working in the areas of materials science and engineering from all over the world. It covered recent trends and progress made in the field of material sciences. Professors from Egypt, UK, Canada, and Spain were invited to deliver keynote lectures regarding the latest information in their respective areas of expertise.

All papers were subjected to peer-review by conference committee members and international reviewers. The papers were selected based on their quality and their relevance to the conference. The proceedings present recent advances in the fields of materials science and engineering and manufacturing as well as in other related areas including materials properties, measuring methods, analytical and numerical modeling and simulations among others. I would like to express special gratitude to members of the conference committee and organizers of the conference. I would also like to thank the reviewers for their valuable time and advice which helped in improving the quality of the papers selected for presentation at the conference and for publication in the proceedings. Finally I want to thank the authors, the members of the organizing committee, the reviewers, the chairpersons, sponsors, and all other conference participants for their support of Eg-MRS 2020.

Prof. Arafa H. Aly

Beni-Suef University

Egypt

IOP Conf. Series: Materials Science and Engineering **956** (2020) 011001 doi:10.1088/1757-899X/956/1/011001

August, 2020

Conference Chairman

Prof. Dr Arafa Hussien Aly
Beni-Suef University, Egypt
Eg-MRS Chiarman

Coordinator (Eg-MRS)

Prof Dr.Hassan Talaat (Ain Shams University, Egypt)
Prof. Dr. Hassan Afify ***(National Research Center, Egypt)***
Prof. Dr.Talaat Meaz ***(Tanta University, Egypt)***
Dr. Amal Kasry ***(The British University in Egypt)***

Editor:Prof Dr. Arafa H.Aly , Beni-Suef Univerrsity, Egypt , E-mail: arafaaly@aucegypt.edu

Reviewers

1. Prof Dr.Hassan Talaat, Ain Shams University, Egypt
2. Prof Dr. Hassan Afify, National Research Center, Egypt
3. Prof Dr.Adel Phillips, Ain Shams University, Egypt
4. Prof Dr.Hamdy Farghaly, El-Minia University, Egypt
5. Prof Dr.Sedky Sodky Abolila , Sohag University, Egypt
6. Prof Dr. Ahmed Ghitas, NRIAG, Helwan, Egypt
7. Prof Arafa H. Aly, Beni Suef University, Egypt
8. Prof Dr. Osama A. Fouad, CMRDI, Egypt
9. Prof Dr. Mona Mohsen, Ain Shams University, Egypt
10. Prof Rania M. Ahmad, Zagazig University, Egypt
11. Prof Dr. Gamal Turky, National Research Center, Egypt
12. Prof Samy Abdel Hkeem, Beni suef University, Egypt
13. Prof. Tarek El-brolossy, Ain Shams University, Egypt
14. Dr.Numan Arshed, Sunway University, China
15. Dr.Afaf ElSayed , Al Azhar University, Egypt

IOP Conf. Series: Materials Science and Engineering **956** (2020) 011002 doi:10.1088/1757-899X/956/1/011002

Peer review declaration

All conference organisers/editors are required to declare details about their peer review. Therefore, please provide the following information:

- **Type of peer review: Single-blind / Double-blind / Triple-blind / Open / Other (please describe) Double-blind**
- **Conference submission management system: Online: https://forms.gle/WksmQNT9sPDkmxQ49**
- **Number of submissions received: 44**
- **Number of submissions sent for review: 25**
- **Number of submissions accepted: 17**
- **Acceptance Rate (Number of Submissions Accepted / Number of Submissions Received X 100):**
- **Average number of reviews per paper: 2**
- **Total number of reviewers involved: 20**
- **Any additional info on review process: Hard work**
- **Contact person for queries: Prof Arafa Hussien Aly**

Please submit this form along with the rest of your files on the submission date written in your publishing agreement.

The information you provide will be published as part of your proceedings.

IOP Conf. Series: Materials Science and Engineering **956** (2020) 012001 doi:10.1088/1757-899X/956/1/012001

Systematic investigation into mechanical strength, pore structure and microstructure of high performance concrete incorporating nano-hybrids

M. Morsy[a], H. Shoukry[a], M.M. Mokhtar[a], N. A. Taha[b], M. S. Morsy[a]

[a] Housing and Building National Research Center (HBRC), Building Physics Institute (BPI), 87 El-Tahrir St., Dokki, P.O. Box 1770, Cairo, Egypt

[b] Housing and Building Research Center, Reinforced Concrete Structure Institute, 87 El-Tahrir St., Dokki, P.O. Box 1770, Cairo, Egypt

Abstract

In this study, high performance concrete (HPC) containing two types of nanoparticles: carbon nano tubes (CNTs) and nano silica (NS) and hybrid nanoparticles (CNTs + NS) were prepared. The compressive strength, phase composition, pore size distribution, surface area, surface morphology and microstructural characteristics were investigated at 7 and 28 days of hydration. In addition, the combined effect of CNTs and silica fume (SF) on the properties of concrete has been investigated in this study for comparison reasons. The results indicated that, all kinds of the utilized nanoparticles provided an enhancement in the compressive strength of the hardened concrete at early and later ages of hydration. Furthermore, Thermogravimetric analysis (TGA) patterns revealed no great difference in the hydration degree of cement with the addition of CNTs. However, Due to its large specific surface area and the pozzolanic activity, CNTs+NS nanohybrid showed the best acceleration efficiency. The concrete containing CNTs+NS nanohybrid exhibited the highest improvement in compressive strength. The pozzolanic reaction induced by NS and silica fume (SF) led to remarkable increase in the surface area and the nanoporosity/gel pores assigned to calcium silicate hydrate (CSH) phase. The SEM analysis showed that, CNTs resulting in bridging nano-size and micro-size pores or gaps within cement hydrates; however, SF and NS are more efficient in filling the pores and gaps between the hydration products and providing more compact microstructure.

Keywords: High performance concrete; Nano-hybrid; Compressive strength; Pore structure; Surface area; Microstructure; TGA

1. Introduction

The production of high-performance concrete can be considered as one of the most important applications of nano materials in construction industry [1]. Nano materials can modify the nano and micro structure of concrete and leads to enhanced mechanical strength at normal curing temperature. Beside the leading role of nano materials in enhancing the mechanical performance, the nano materials considerably improve volume stability, durability, and sustainability of cement and concrete structures [2]. Application of nano technology in concrete has been implemented by using a wide range of nano materials with various chemical and physical characteristics including: Spherical nano particles (like; nano silica, nano alumina, calcium carbonate nano particles, etc.); nanotubes (like; CNTs) or as nano platelets (like; nano clay and nano graphene) [3]. The role of nano pozzolan like NS in improving concrete performance differs from the role of nano fibers like CNTs. NS possesses chemical and physical effects: (1) pozzolanic reaction; i.e., it reacts with the calcium hydroxide which liberates upon hydration of cement and produces supplemental calcium silicate hydrate which contributes to strength development; (2) Due to its ultra-fineness, NS will realize significant reduction in the amount of micro pores by inducing physical packing, providing a more dense concrete [4,5]. CNTs don't have any chemical affinity with the cement hydrates [6] thereby CNTs are used to reinforce cementitious matrix and attain resistance to micro-crack propagation [7-10]. The effect of NS with mean particle size of 120 nm on the hydration and microstructure characteristics of ultra-high performance concrete (UHPC) with a low cement content has been studied by R. Yu et al. [11] an enhancement by about 28 % was achieved for the 28th day compressive strength of concrete incorporating 4 wt.% of NS. A.M. Said et al. [4] examined the characteristics of concrete containing smaller size NS particles of about 35 nm. The results revealed an enhancement in the 28th day compressive strength by about 36% with the loading of NS up to 6%. Nazari and Riahi [12] investigated the influence of extremely smaller/ultrafine NS of mean particle size of about 15 nm on the compressive strength of concrete, 70% improvement can be achieved with the addition of NS up to 5 wt.%. From the literature survey conducted as a part of the current study, it was concluded that the enhancement ratios in the compressive strength in addition to the optimum NS content are strongly related to the mean particle size of the used NS. The feasibility of CNTs as reinforcement component is not only due to their exceptional high modulus, but also to their high length to diameter aspect ratio. Indeed, the higher the aspect ratio of CNT the higher the contribution to the reinforcement efficiency [13]. Madhavi et al. [1] presented a study on the nano-modification of concrete by multi wall carbon nano tubes (MWCNTs), concrete samples have been prepared in which MWCNTs with diameter < 40 nm and length < 10 µm were added with the aid of sonication at various ratios ranging from 0.015 and 0.045 wt. %. The highest enhancements obtained for the compressive and splitting tensile strengths were 26.69% and 66.3%, respectively at 0.045% MWCNTs addition. Due to the modest contribution of CNTs to compressive strength as reported in many studies [1, 14, 15] a limited number of researchers tried to add/utilize another micro or nano-pozzolanic materials and investigate their hybrid effect with CNTs to attain additional enhancements. Kim et al. [16] tried to increase the 14 days compressive strength of cement mortars incorporating 0.15 and 0.3 wt. % MWCNTs by adding 10, 20 and 30 wt. % SF, The addition of SF showed promising enhancement in the compressive strength. The concrete mix incorporating 10 % SF and 0.15 % MWCNTs showed the highest compressive strength enhancement i.e., this mix attained 1.29 times higher strength than that of the plain concrete (without SF). Gurumurthy et al. [17] added 1 wt. % nano alumina (NA) into cement composites incorporating 0.75 wt. % MWCNTs, The NA provided additional enhancements in the 28th day compressive strength, flexural strength and toughness by about 27.53%, 30.22% and 98.69%, respectively. Hunashyal et al. [18] added 0.5 wt. % NS to cement pastes containing 0.75 wt. % MWCNTs. The 28th day compressive strength, flexural strength, toughness index and ductility index have been improved by about 12.6%, 51.15%, 72.6% and 5.4%, respectively. Most recent studies on supplementary cementitious materials (SCMs) has concentrated on: examining new materials, changing replacement ratios, adapting new test methods and treating or activating materials to improve performance; nevertheless, limited studies dealt with advances in understanding the mechanisms of improving concrete performance by SCMs. The objective of the current study is to investigate more deeply the role of micro- and nano-supplementary cementitious materials (SF & NS) in enhancing properties of CNTs- modified concrete.

2. Experimental details:

2.1. Materials

2.1.1. Cement

Ordinary Portland cement (OPC: CEM I 42.5 R) was obtained from Suez- Factory in Egypt.

2.1.2. Coarse aggregate

The Corse aggregate was crushed dolomite with nominal maximum size of 20 mm obtained from Attaka quarry.

2.1.3 Fine aggregate

IOP Conf. Series: Materials Science and Engineering **956** (2020) 012001 doi:10.1088/1757-899X/956/1/012001

Siliceous sand was used as a fine aggregate with particle size ranging from 0.06 to 4.76 mm.

2.1.4. Water reducer Superplasticiser

Sikament 163M was used as water reducer. Sikamaint 163 has been added as plasticizer with dosage 1.5% of cement weight was incorporated into all mixes to ensure the good workability and no segregation would occur.

2.1.5. Silica fume (SF)

SF with specific surface area of 17 m^2/g was supplied from an Egyptian company making ferrosilicon alloys (EDFO). The SF particles are spherical in shape with average diameter of between 1–5 μm.

2.1.6. Nano silica (NS)

NS synthesized by sol-gel procedure were obtained from housing and building National research center (HBRC). The average primary particle size is of about 20-30 nm with specific surface area of about 230 m^2/g.

2.1.7. Multiwall carbon nano tubes (MWCNTs)

MWCNTs synthesized by atmospheric pressure chemical vapor deposition (APCVD) were obtained from housing and building National research center (HBRC). The outer diameter is about 10-25 nm; however, the length is of about 25μm.

2.2. Samples preparation

Concrete cubes were prepared with a binder content of 350 kg/m^3. The blended concrete samples were prepared in which SF and NS replacing 7.5 % and 2 % by weight of cement respectively. In order to develop the nano hybrid-modified concrete mixes, MWCNTs were added by 0.03 wt. %. All mixes were prepared with a constant water/ binder (w/b) ratio. Table 1 illustrates the mix design of the plain and nano-modified concrete. During the preparation of nano modified concrete, the sonication method has been considered to avoid the particle fusion/segregation and to attain high degree of dispersion of nano additives in the blended concrete. NS and CNTs were first added to the mixing water dosed with superplasticiser, the mixtures were stirred with an ultrasonic emitter for 30 min to form an aqueous solution. The aqueous solution with the dispersed/suspended nano particles was mixed with the cement and aggregates using a laboratory electric mixer to prepare the blended concrete. Along with the concrete samples; cement pastes incorporating the nano hybrids have been prepared, especially for the TGA, BET and XRD analyses in order to diminish the effect of aggregates on the accuracy of analyses.

Table (1): Concrete mix proportions (kg/m^3)

Mix ID	OPC	SF (replacement)	NS (replacement)	CNTs (addition)	Gravel	Fine Sand	Water	Super plasticizer
Control	350	--	--	--	1470	435	125	6
CNT	350	--	--	0.105	1470	435	125	6
CNT + NS	343	--	7	0.105	1470	435	125	6
CNT+SF	323.75	26.25	--	0.105	1470	435	125	6

2.3. Test methods

2.3.1. Compressive strength

The compressive strength test was conducted on 15 cm cubic concrete specimens at the 7th and 28th days of curing in accordance with B.S. 1881-116, [19], using 2000 KN Italy Matest pressing machine. The loading rate of 6.5 KN/min was considered.

Three specimens were tested from each mix and the average was calculated.

The pressure resistance of each sample was calculated as follows:

$$\textbf{Compressive strength} = \frac{\text{load in "N"}}{\text{Area in "mm}^2\text{"}}; \textbf{MPa} \qquad (1)$$

IOP Conf. Series: Materials Science and Engineering **956** (2020) 012001 doi:10.1088/1757-899X/956/1/012001

2.3.2. Surface area and pore structure analysis

Brunauer-Emmett-Teller (BET) method was used to assess the specific surface area (SSA) of the hardened pastes and the Nitrogen adsorption–desorption isotherms were collected using a NOVA2000e (Quantachrome, USA) instrument. Barrett–Joyner–Halenda (BJH) method was adapted for the pore size distribution patterns.

2.3.3. X-ray diffraction analysis

The XRD analysis was carried out to identify the phase composition using a PW 1050/70 Diffracometer (Philips, Holland), equipped with monochromatic Cu-Kα radiation. XRD patterns were recorded in the 2θ range from 5° to 50° at a scanning rate of 2°/min. This test was performed on fine grinded samples passing 75μm sieve to ensure homogenous phase characterization.

2.3.4. Thermogravimetric analyses (TGA)

DSC-TGA Thermogravimetric analyzer (STD-Q600, USA) was used to explore the phase composition of the nano-hybrid modified concrete.

2.3.5. Scanning Electron Microscope (SEM)

The microstructure characteristics were investigated by environmental scanning electron microscope (ESEM- Inspect S, FEI Company, Holland). The 3d-surface topography analysis was performed by the *image J* software.

3. Results and discussions

3.1. Compressive strength

Figure (1) presents the development in compressive strength for the concrete containing CNTs and the hybrid systems (CNT+NS) and (CNT+SF) at 7and 28 days of hydration.

As it was expected, the blended concrete showed improved compressive strength at even at early or later ages of hydration. Table (2) summarizes the enhancement ratios achieved after 7 and 28 days for the developed concrete blends.

Table (2): Results for compressive strength tests and the enhancement ratios at 7 and 28 days of hydration

Mix	Compressive strength, MPa		Enhancement ratio, %	
	7 d	28 d	7 d	28 d
Control	26.1	36.78	--	--
CNTs	34.93	41.80	33.83	13.37
CNTs+NS	36.47	50.23	39.73	36.23
CNTs+SF	32.77	46.8	25.55	26.93

Although, the CNTs showed a significant enhancement at 7 days of curing, the concrete incorporating CNTs + NS showed the highest enhancement ratios. Actually, the high aspect ratio of CNTs is a very important advantage that makes CNTs contribute physically in improving the mechanical performance and microstructure of concrete, thus serve as a fibers forming crosslinks between cement hydrates, resulting in arresting and bridging micro cracks inside the hardened structure and therefore could suppress crack propagation in the matrix at nanoscale before being pulled out or stressed to rupture [21, 22]. The integration of NS with CNTs, improves the hydration acceleration efficiency and contributes to the early age strength, this is attributed to its nanosize which allows NS to actively act as an active filler thereby providing nucleation/seed centers for the cement hydrates [23]. The combination between CNTs and SF is found to be not effective at early ages and reduces the gain of strength in the CNTs-modified concrete; this is may be assigned to the agglomeration of SF around CNTs within the cementitious structure skeleton which adversely reduces the acceleration efficiency of the cement hydration. After 28 days of hydration, CNTs provided 13.37 % an enhancement in compressive strength relative to the normal/plain concrete; nevertheless, the coupling with SF and NS provided an additional enhancement by about 14% and 23% respectively. These additional enhancements are resulted from the pozzolanic activities of SF

IOP Conf. Series: Materials Science and Engineering **956** (2020) 012001 doi:10.1088/1757-899X/956/1/012001

and NS. However, the variation in the provided enhancement ratios is related to the grain size distribution, specific surface area and chemical activity of SF and NS. While NS being nanostructured and composed mainly of active amorphous SiO_2 (i.e. more active than SF), these features allow it to play a vital important role in improving mechanical strength of concrete; physically through filling the pores, and chemically through its high pozzolanic activity. In this process SiO_2 reacts with the harmful hydration product- $Ca(OH)_2$- producing more calcium silicate hydrate phase in the concrete skeleton and providing a quite compact, homogenous and more dense structure [24,25].

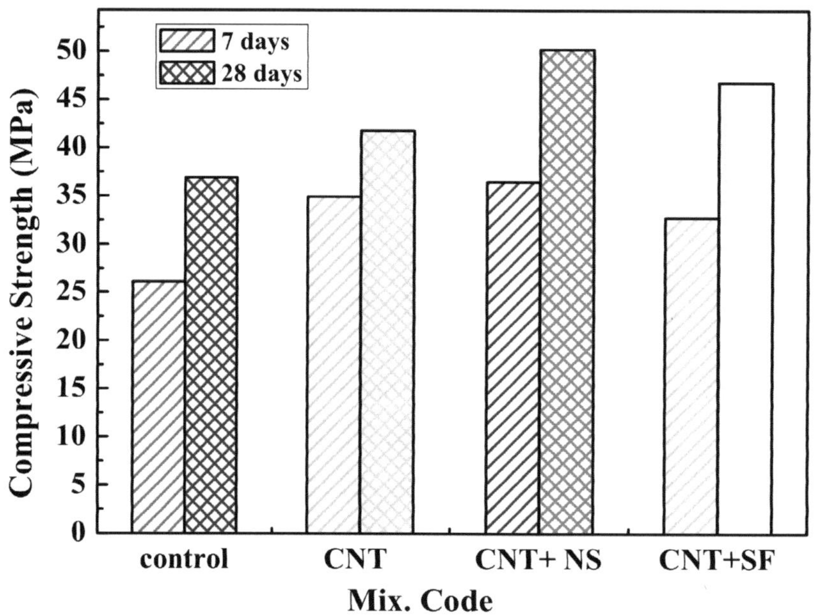

Fig. 1: Compressive strength results of concrete reinforced with CNTs, CNTs + NS and CNTs + SF in comparison with normal concrete.

3.2. Surface area and pore structure

N_2-adsorption/desorption isotherms and BJH pore size distribution of hardened cement pastes incorporating CNTs and the hybrid systems (CNTs + NS) and (CNTs + SF) compared to the plain ordinary cement after 28d of curing is presented in fig.2.

As it is clear from the adsorption/desorption isotherms, the examined pastes showed considerable variation in the quantity of the adsorbed nitrogen, depending dramatically on the pore size characteristics and specific surface area (SSA). The four pastes exhibited a physisorption isotherm of type IV (i.e. porous materials), as evidenced by IUPAC classification [26]. Both the control paste and the paste containing only CNTs adsorbed approximately the same quantity of N_2 manifesting that, they are exhibiting the same surface area. The pastes incorporating the nano hybrids adsorbed more nitrogen due to their significant higher SSA. The SSA of the control, CNTs, (CNTs+SF) and (CNTs+NS) modified pastes are 6.78, 7.06, 13.65 and 17.46 m^2/g respectively. The increase in the SSA with nano hybrids addition is logical as the possibly reactive area is increased due to the addition of extra reactive surfaces. This finding conform with the results of previous study which assigned the role of fine minerals including SF, FA, and fine quartz in improving the hydration acceleration efficiency of cement to the increased surface area for CSH nucleation/growth [27]. Furthermore, this may explains the additional enhancements achieved in the mechanical strength of the concrete incorporating hybrid systems and testifies the hydration acceleration provided by NS due to its large SSA and the pozzolanic activity [28]. According to their sizes, pores are categorized into four groups; large pores (macropores) of size >1000 nm, capillary pores have sizes range from 100 to 1000 nm, while transitional or meso-pores are of size between 10 and 100nm, and finally gel pores or nanopores which is considered to be the pores between the (CSH) gel layers have sizes <10 nm [29]. The pore size of the test samples is mainly around 1.5 nm, As it can be observed from the BJH curves, the pores in our prepared mixes are mainly around 1.5 nm; i.e., located in the range of gel pores. It is also clear that, CNTs have not significant effect on the pore volume as it is chemically inert with the cement hydrates. The hybrid systems have a great effect on modifying the pore structure, where the total volume of gel pores was markedly increased. Although the hybrid system (CNTs + SF) showed a considerable increase in the pore volume, the nano hybrid (CNTs + NS) showed the highest intensity due to the higher pozzolanic activity of the nano scale SiO_2. The pore volume of the control, CNTs, (CNTs + SF) and (CNTs + NS) modified pastes are 0.014, 0.018, 0.024 and 0.03 cm^3/g respectively. The increased volume of gel pores with the addition of SF and NS is mainly assigned to the filling effect and the formation of additional C–S–H through C2S hydration and pozzolanic reaction [30–32].

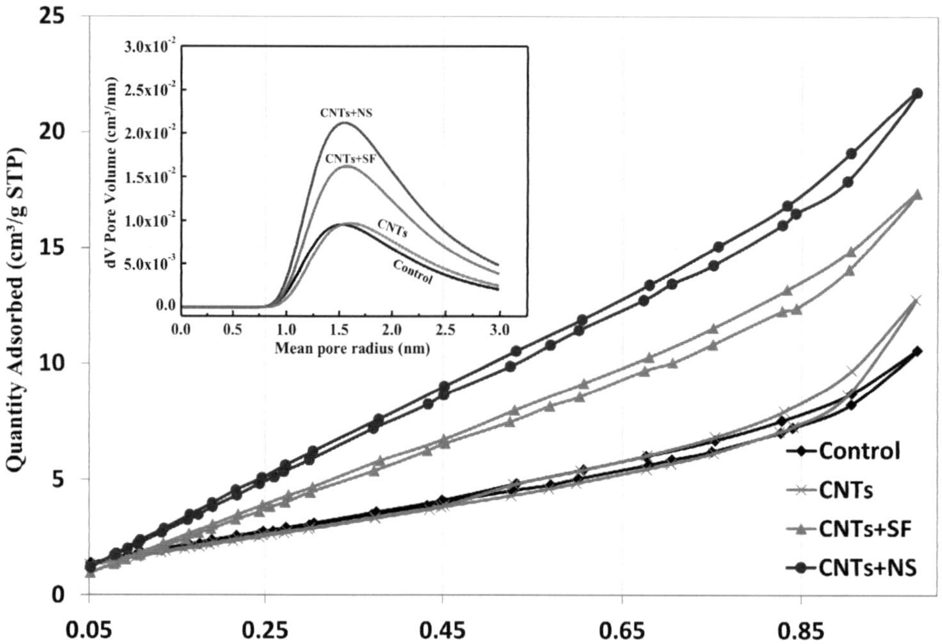

Fig.2: N2-adsorption/desorption isotherms and BJH pore size distribution of cement pastes incorporating nano hybrids at 28 days of curing.

3.3. Phase composition analysis by XRD

The XRD patterns of the cement pastes incorporating CNTs, (CNTs + SF) and (CNTs + NS) as compared with the control reference paste at 7 and 28 days of curing are shown in fig.3. Tricalcium silicate (C3S), dicalcium silicate (C2S), calcium hydroxide (CH), CSH, and calcium carbonate (CC) are the main phases detected. Evidently, the intensities of cement hydrates including CSH and CH increases with the curing time, whereas the peaks of the un-hydrated OPC phases (C3S and C2S) decrease with curing time due to the progress of hydration process. By comparing the CH peaks of the control paste and the paste incorporating only CNTs even at 7 and 28 days of hydration, it is clear that the intensity of CH peaks did not change due to the addition of CNTs. this behavior confirm the very limited role of CNTs in accelerating the hydration kinetics. These results comply with that of TGA and BET. While for cement modified with combination of (CNTs + NS) and (CNTs + SF); the CSH peaks were observed to increase at the expense of CH peaks as a result of the pozzolanic reaction between SiO_2 contained in NS and SF with CH, and this explain the remarkably enhanced mechanical performance of (CNT_S + NS) and (CNTs + SF) reinforced concretes compared to the ordinary concrete. Reducing the CH content in concrete contributes to enhanced strength and improved durability characteristics of concrete structures [33].

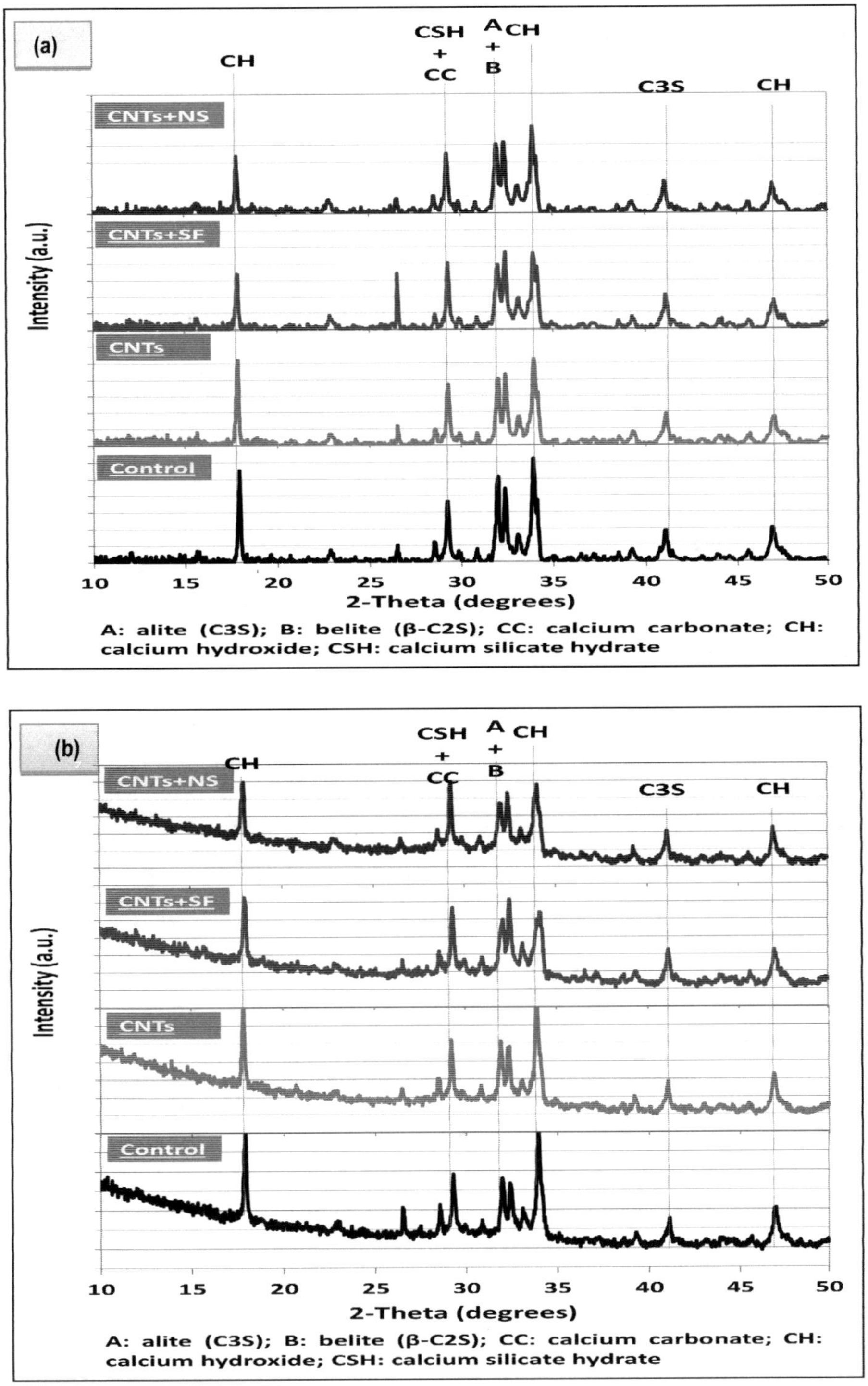

Fig.3. XRD patterns of cement pastes modified with nano hybrid systems hydrated for (a) 7 days and (b) 28 days.

3.4. Thermo gravimetric analysis (TGA)

The TGA results of the cement pastes incorporating CNTs, (CNTs + SF) and (CNTs + NS) as compared with the control reference paste hydrated for 7 and 28 days are displayed in figures (4 & 5) respectively. The decomposition temperatures of the hydration products are the same at both 7 and 28 days of hydration, there were a mass loss at about 200 ^{o}C which is due to the CSH phase decomposition, while the reduction in mass at about 470 ^{o}C is assigned to the decomposition of CH phase, furthermore, the mass loss appeared at about 700°C is attributed to crystalline $CaCO_3$ decomposition [3]. From the obtained thermograms, it can be noted that; at 7 days of curing, both control concrete and concrete reinforced with CNTs have almost identical behavior and values, this means that, they have the same CSH and CH contents, where CNTs have no role in promoting the hydration process of cement. On the other hand, the addition of a pozzolanic material to concrete has modified the hydration process resulting in an increase in CSH content and the CH content decreased for concrete reinforced with combinations of (CNTs + NS) and (CNTS + SF), respectively. Furthermore at 28 days of curing; the hydration reaction is almost completed and the hydration phases are formed. Again as it was observed at 7days, both normal concrete and concrete reinforced with only CNTs have the same hydration products content because of, there is no pozzolanic reaction took place, where CNTs has no chemical contribution in the hydration reaction. They have the highest CH content and the lowest CSH content.

While the lowest CH content and the highest CSH content was found in the concrete-CNTs/NS mix followed by concrete-CNTs/SF mix., which confirms the high pozzolanic activity of both NS and SF, which in turn reflected on the mechanical performance of concrete. This explained behavior was also proofed by the XRD patterns.

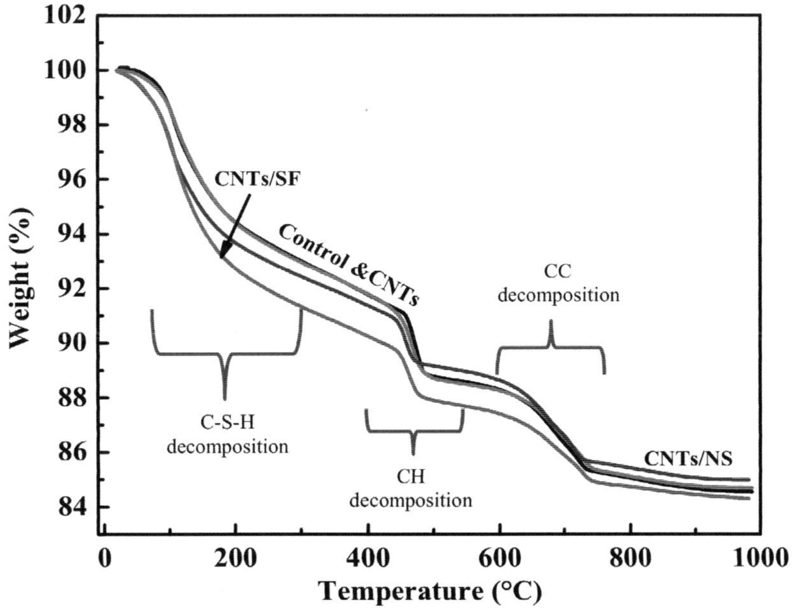

Fig. 4: TGA Thermograms of concrete reinforced with CNTs, CNTs/NS and CNTs/SF compared to normal concrete, at 7 days of curing.

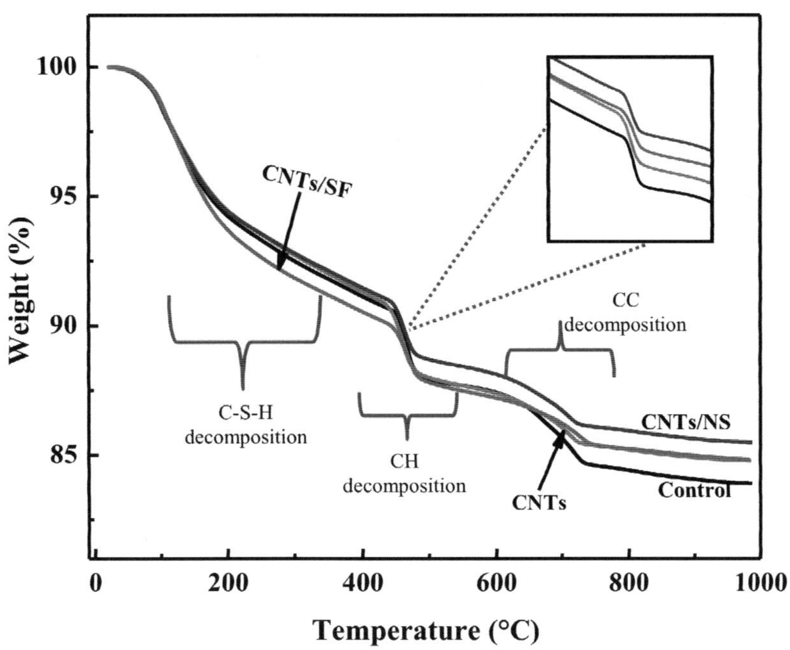

Fig. 5: TGA thermograms of concrete reinforced with CNTs, CNTs/NS and CNTs/SF compared to normal concrete, at 28 days of curing.

3.5. Microstructure characteristics

Figure 6 presents the SEM micrographs along with the surface topography/roughness analysis of the control, CNTs, (CNTs + SF) and (CNTs + NS)-modified pastes at 28 days of curing. The microstructure of both the control and CNTs-blended pastes showed the presence of hydration products including; gel-like CSH, hexagonal CH crystals with micro cracks and micro pores. The crack bridging provided by CNTs can be clearly seen. With the inclusion of SF and NS, the gel-like CSH is the predominant phase with the absence of large CH crystals; furthermore, the actions of micro filler and physical packing are evident. Although the surface topography analysis for all cement based mixtures are classified as rough and irregular surfaces. The mixture containing (CNTs + NS) exhibited relatively smoother surface, this is attributed to the unique filler and pozzolanic actions of the nano hybrids and points for the increased gel-like CSH content. This result is compatible with the findings of a previous study, which assigned the smooth surface characteristic of UHPC containing nano metakaolin to the pozzolanic reaction caused by metakaolin [34].

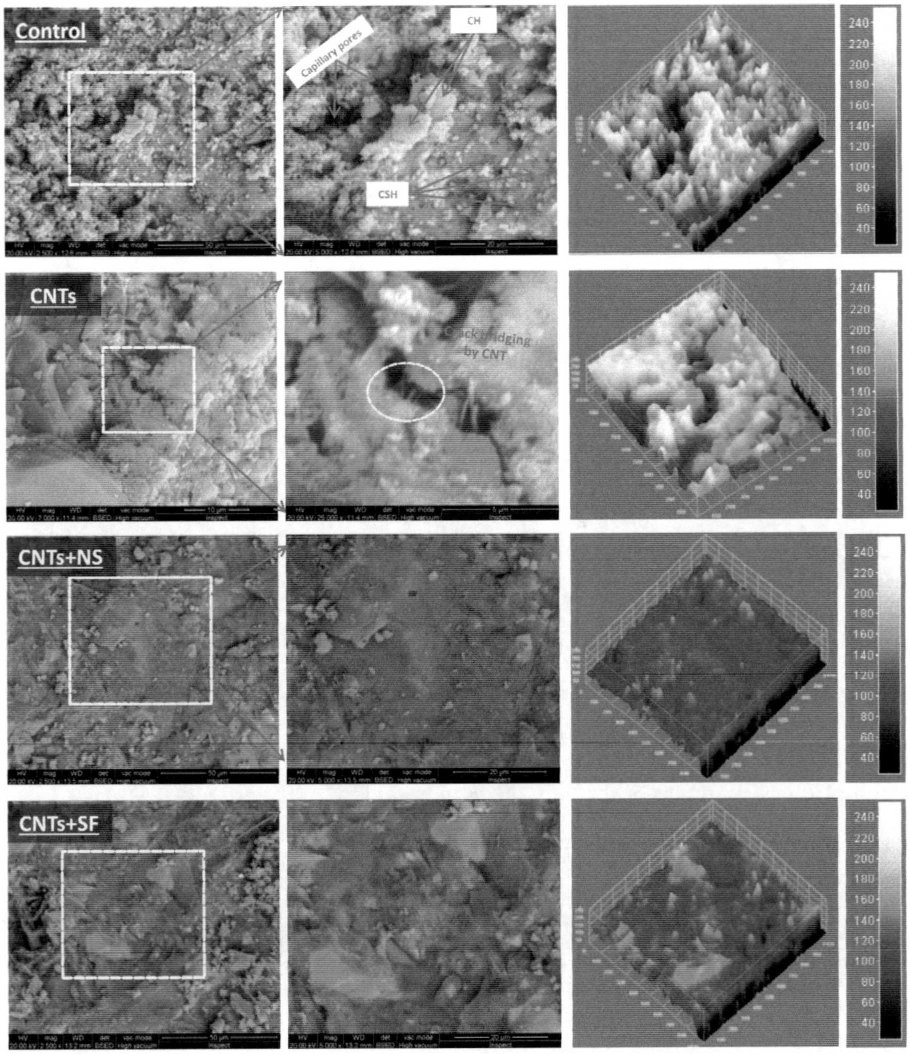

Fig. 6: SEM micrographs and surface topography analysis of cement pastes modified with nano hybrid systems hydrated for 28 days.

IOP Conf. Series: Materials Science and Engineering **956** (2020) 012001 doi:10.1088/1757-899X/956/1/012001

4. Conclusions

In this work the combined/hybrid effect of micro- and nano-supplementary cementitious materials including (SF & NS) along with CNTs on the HPC have been investigated. The compressive strength, phase composition, BJH pore size distribution, BET surface area and microstructural characteristics have been determined; the experimental results testify the following conclusions:

- Owing to their fiber like morphologies with very high aspect ratios, CNTs contribute in enhancing the mechanical performance by improving the microstructure of concrete. CNT_S are helpful in forming crosslinks between cement hydrates leading to arresting and bridging of micro cracks in the matrix at nanoscale.
- CNTs don't provide significant acceleration effect on hydration degree of concrete.
- The integration of NS with CNTs enhances the hydration acceleration efficiency and contributes to the early age strength. This is attributed to its nano spherical morphology which allows NS to actively act as active filler thereby providing nucleation centers for the cement hydrates.
- The integration of nano hybrids into the concrete led to significant increase in the BET surface area this is ascribed to the addition of extra reactive surfaces.
- The hybrid systems (CNTs + NS) and (CNTs + SF) are effective in providing additional enhancement in the compressive strength and contribute to improved microstructure. This is attributed to the unique filler and pozzolanic actions of the nano hybrids that result in the increased gel-like CSH content.
- As it is chemically inert with the cement hydrates, CNTs have insignificant effect on the pore volume; however, the hybrid systems have a great effect on modifying the pore structure.

5. References

[1] Madhavi T C H, Pavithra P, Sushmita B S, Vamsi R S B, and Surajit P 2013 *Engineering* **2(6)** 166–168.

[2] Grigorij Y, Jadvyga K, Albinas G, and Ingrida G, 2006 *Material Science* **12(2)** 147–151.

[3] Morsy M S, Shoukry H, Mokhtar M M, Ali A M, and El-Khodary S A, 2018 *Construction and Building Materials*, **172** 243–250.

[4] Said A M, Zeidan M S, Bassuoni M T, and Tian Y, 2012 *Construction and Building Materials* **36** 838-844.

[5] Juenger M C G, and Siddique R, 2015 *Cement and concrete research* **78** 71-80.

[6] Markar J M., Margeson J C, and Luh J, NRC Publications Record, 3rd International Conference on Construction Materails: Performaing Innovations and Structural Implications, Vancouver, Aug 22-24, 2005, pp. 1–10.

[7] Kang S T, Seo J Y, and Park S H, 2015 *Advances in materials science and engineering* **2015(6)** 1–9.

[8] Xu S, Liu J, and Li Q, 2015 *Construction and Building Materials* **76** 16–23.

[9] Manzur T, Yazdani N, and Emon M A B, 2014 *Journal of Materials*, **2014(1)** 1–8.

[10] Zhou X , Shin E, K Wang, and Bakis C, 2004 *Composites Science and Technology* **64(15)** 2425–2437.

[11] Yu R, Spiesz P, and Brouwers H J H, 2014 *Construction and Building Materials* **65** 140–150.

[12] Nazari A, and Riahi S, 2011 *Composites Engineering* **42(3)** 570–578.

[13] Abu Al-Rub R K, Ashour A I, and Tyson M B, 2012 *Construction and Building Materials*, **35** 647-655.

[14] Kay W and Kenneth J L, Transportation Research Board of the National Academies, Washington, 2142, 2010, pp. 119–126.

[15] Camacho M D C, Galao O, Baeza F J, Zornoza E, and Garcés P, 2014 *Materials* **7** 1640-1651.

[16] Kim H K, Nam I W, and Lee H K, 2014 *Composite Structure* **107** 60–69.

[17] Gurumurthy G, Hunsahyal A M, Quadri S S, Banapurmath N R, Shetter A S, Hiremath R K, and Hallad S A, 2014 *International Journal of Advance Research In Science And Engineering* **3(8)** 215–223.

[18] Anand H, Nagaraj B, Akshay J, Sayed Q, and Ashok S, 2014 *Advanced Materials*, **3(5)** 45–51.

[19] B.S. 1881, Part 116, 1989, Method for determination of compressive strength of concrete cubes, British Standards Institution.

[20] Siddique R, and Mehta A, 2014 *Construction and Building Materials* **50** 116–129.

[21] Makar J M, and Chan G W, 2009 *Journal of the American Ceramic Society*, **92(6)** 1303–1310.

[22] Gamal H A, El-feky M S, kohail M, and Nasr E A R, 2018 fifteenth international conference on structural and geotechnical engineering, Advancements in Construction Techniques.

[23] Kumar S, Kumar A and Kujur J, Proceedings of the Institution of Civil Engineers – Structures and Buildings, https://doi.org/10.1680/jstbu.18.00080

[24] Wan B J, Hyun K C, Ghi-ho T and Jang-Bin P, 2007 *Construction and Building Materials* **21** 1351-1355.

[25] Morsy M S, Alsayed S H, and Aqel M, 2011 *Construction and Building Materials* **25** 145–149.

[26] El-Feky M S , El-Khodary S A, and Morsy M, 2019 *Egyptian Journal of Chemistry* **62** 57-67.

[27] Aditya K, Tandre O, and Gabriel F, 2017 *Journal of Ameriacan Ceramic Society* **100** 1-13.

[28] Land G, and Stephan D, 2012 *Journal of Materials Science* **47** 1011-1017.

[29] Mokhtar M M, Abo-El-Enein S A, Hassaan M Y, Morsy M S, and Khalil M H 2017 2017 *Construction and Building Materials* **138** 333–339.

[30] Hou P, Kawashima S, Kong D, Corr D J, Qian J, and Shah S P, 2013 *Composite B Engineering* **45(1)** 440–448.

[31] Zapata L E, Portela G, Suárez O M, and Carrasquillo O, 2013 *Construction and Building Materials* **41(41)** 708–716.

[32] Choolaei M, Rashidi A M, Ardjmand M, Yadegari A, and Soltanian H, 2012 *Materials Science and Engineering* **A538 (3)** 288–294.

[33] Al-Salami A E, Morsy M S, Taha S, and Shoukry H, 2013 *Construction and Building Materials* **47** 138–145.

[34] M.S. M. Norhasri, M.S.Hamidah, and A. Mohd Fadzil, "Inclusion of nano metaclayed as additive in ultra high performance concrete (UHPC)", Construction and Building Materials 201 (2019) 590–598.

Characterization and Dielectric Properties of Fly Ash / Polystyrene / Low Density Polyethylene Composites

G M Nasr[a], T A Mohamed[b,c], R M Ahmed[b,*]

[a] *Physics department, Faculty of Science, Cairo University, Giza, Egypt.*
[b] *Physics department , Faculty of Science, Zagazig University, Zagazig 44519, Egypt.*
[c] *Zagazig Higher Institute of Engineering and Technology, Zagazig, Egypt.*

* E-mail of the corresponding author: rmkhalil@zu.edu.eg

Abstract: Polystyrene (PS) and its blend with low density polyethylene (LDPE) were prepared by melting method. Fly ash was used as filler with different contents to form composites by the same way with PS/LDPE 80/20 blend. The obtained data of fly ash from X–Ray Fluorescence analysis displayed that the total mass summation of silicon, magnetite and aluminum (SiO_2 + Fe_2O_3+ Al_2O_3) is about 79.52 wt. % whereas the amount of calcium oxide is less than 10%. TGA curves showed that the sample of the more thermal stability was the composite that contains the highest amount of fly ash (50 wt.%). For all the samples, the dielectric constant and the dielectric loss were increasing by decreasing the applied frequencies while they were decreasing by increasing the temperatures. Also, increasing the fly ash content in the blend improved the ac conductivity up to a content of 40 wt.% then it decreased. The values of the real part of the electrical modulus (M') did not equal to zero at low frequencies and high temperature. Therefore, it is expected that the electrode polarization may take place for the all samples.

1-Introduction

Few decades ago, polymers and polymer-based materials have been used for various electrical applications. Numerous studies have concerned on the improvement of the polymeric material with low and high dielectric constant. For instance, polymers with high dielectric constant have the ability to be applied as artificial muscle and film capacitors. However, those with low dielectric constants can be utilized in the field of capacitors and dielectric materials [1-3]. Polystyrene (PS) is a commercial thermoplastic polymer. It is rather brittle, clear and has good mechanical properties and a low cost price. Thus, PS has a wide range of applications as construction materials, packaging, disposable cups, consumer electronics, cassette boxes, compact disks and medical uses [4, 5]. This study includes also another polymer which is the low density polyethylene (LDPE). It is nominated to be widely used in different applications as using it in the industries of cables and wires. This is a consequence of its high resistivity and breakdown strength. LDPE was used in some other applications as packaging film, decks and docks, window frames, tubes and also in the industry of the automobile [6].

A great interest of research was paid for the polymer matrix composite which has unusual modified physical properties and variant industrial applications [7-10].The physical/chemical properties of the hosting polymer and the filler can affect greatly the properties of their composites [11]. Nowadays, a great attention has been paid for developing the electrical conductivity and the dielectric properties of polymer composites. Therefore, variant kinds of fillers have been employed for improving the conductive ability of the polymeric materials. Developing the physical properties of the polymers is produced from the amalgamation of fillers with them [12].

In fact, using fillers is preferable for variant applications in order to lower the high cost of some hosting polymers and modifying their properties [6]. The effects of fly ash, as fillers, on the physical properties of the polymer materials were examined widely in many literatures [13-16]. Day by day, the production of fly ash has increased which is considered as a waste that is formed from the thermal power plants. Fly ash has become one of the most popular wastes which are used as industrial raw material. The production of fly ash resulted in many environmental and economical problems. Therefore, the researchers work on recycling it in different industrial aspects [17]. Fly ash has possibly included in diverse applications as using it in manufacturing of ash bricks, metal recovery, water and atmospheric pollution control and also in agriculture [18]. Fly ash contents considerable amounts of silica (SiO_2), alumina (Al_2O_3), iron oxide (Fe_2O_3), calcium oxide (CaO). However, magnesium oxide (MgO), titanium oxide (TiO_2), alkaline (Na_2O and K_2O), sulfur trioxide (SO_3), phosphorus oxide (P_2O_5) and carbon have usually small percentages [19].

As mentioned previously, fly ash is considered as a waste material which has a negative impact on the environment. Therefore, in the present study, the authors tried to reuse this low cost inorganic waste by embedding it in polymeric blends and studying its effect on the physical properties of the host matrix. PS, PS/LDPE blend and composites of fly ash/PS/LDPE were prepared by melt mixing method. The effect of fly ash as filler on tuning the electrical properties of the hosting blend of PS/LDPE was investigated. The fly ash was characterized by XRF analysis. The thermal stability of the samples was examined by TGA analysis. AC conductivity, dielectric constant, dielectric loss and electrical modulus were deduced and analyzed at a wide range of applied frequencies and variant constant temperatures.

2-Experimental Techniques

2.1 Material and the Samples

For this study, polystyrene (PS) was purchased from Sigma Aldrich (Germany) and low density polyethylene (LDPE) was obtained from advent chembio (India). Fly ash (FA) powder was supplied from the Nile Sons Co. for brick manufacturing (Cairo, Egypt) and it was used without modifications.

The prepared samples included pristine PS and a blend of PS/LDPE 80/20 wt. % which labeled as PS and C0, respectively. Also, Fly ash was added to PS/LDPE 80/20 blend to form composites with different concentrations. The fly ash contents included 10, 20, 30, 40, 50 wt.% and have labels of C10, C20, C30, C40 and C50, respectively. The samples were prepared by direct mixing in an internal mixer (Bra-bender plasticorder, USA) at a temperature of 90 ^{o}C for LDPE and 215 ^{o}C for the others. The process of mixture was done with a rotor speed of 80 rpm for 10 min. By using the electrically hydraulic press (XLB-D vulcanizing press, Zhejiang Hongtu machinery factory, China), the melt-blended samples and their composites were compression molded. This was done at 190°C under a constant pressure of (13.5 Mpa) followed by air cooling. Each sample was placed in a stainless steel mold of dimension 100 x100 x 2 mm^3 to be compressed and molded between the two plates of the hydraulic press.

2.2 XRF Measurements

The chemical composition of fly ash was determined by using wavelength dispersive X–Ray Fluorescence (XRF, PANalytical, Axios mAX, Netherlands).

2.3 DTA and TGA Measurements

Thermal gravimetric analysis (TGA) and differential thermal analysis (DTA) were used to determine the thermal stability of the tested material. TGA was performed, using Shimadzu Thermobalance model DTG-60H, for the pristine PS, PS/LDPE blend, C10 and C50. The tested samples were heated under nitrogen gas atmosphere with a heating rate of 5 ^{o}C/minute started from the room temperature to 600 ^{o}C. The initial weight of the samples was 7.1 ± 1.48 gram for each one.

2.4 Dielectric Measurements

Ac conductivity and the dielectric properties were examined for the whole samples in the form of circular discs, with a diameter of 1 cm. For good conduction, silver past was used to coat opposite surfaces of each sample. The measurements were done in temperature range of 303 - 403 K at frequencies from 1 to 5000 kHz. RLC Bridge (IM 3570 impedance analyzer "Hioki, Japan") was used to investigate these measurements.

The real part (ε', dielectric constant) and the imaginary part (ε'', dielectric loss) of the relative complex permittivity ε^*, of a material is given by: $\varepsilon^* = \varepsilon' - j\varepsilon''$. The dielectric constant ε′ describes the amount of both induced and permanent dipole alignment. $\varepsilon' = C/C_o$; where C and C_o is the sample and vacuum capacitance, respectively. The relation of $\frac{\varepsilon_o \times a}{d}$, describes the value of C_o where ε_o is the free space permittivity (ε_o = 8.86 × 10^{-12} F/m), *d* is the sample thickness and *a* is the effective electrode. The dielectric loss ε″ gives the energy required to align dipoles or move ions. $\varepsilon'' = \varepsilon' \tan\delta$, where tanδ is the dielectric loss tangent [7, 20]. Ac conductivity, σ_{ac}, was calculated from dielectric data through the relation of $\sigma_{ac} = \varepsilon''\varepsilon_o\omega$, where ω is the angular frequency and $\omega = 2\pi f$ [21].

3- Results and Discussion

3.1 XRF Analysis

Table 1 illustrates the composition of fly ash that was analyzed with X-ray fluorescence (XRF). Iron, calcium, silicon, magnesium, aluminum, and sulfur oxides are some minerals of the chemical composition of fly ash. Some other elements in the ash possess high melting points which remain in furnace during combustion process. Magnetite (Fe_2O_3), lime (CaO), and periclase (MgO) are formed from oxygen in the air combined with Fe, Ca, and Mg. The fly ash is divided into two classes, depending on American Society for Testing and Materials (ASTM) standard: class C and class F. These classes are mainly relying on the chemical composition of the material as the amounts of the oxides including silicon, alumina and iron oxides. Nevertheless, the major dissimilarity between class C and class F ashes is the sum of the oxides

(minimum 70% for class F ashes and 50% for class C ashes) [22]. In this study, the total mass summation of silicon, magnetite and aluminum (SiO_2 + Fe_2O_3+ Al_2O_3) is about 79.52 wt. % whereas the amount of calcium oxide (~ 9.5) is less than 10%. Consequently, the used fly ash in this study can be considered according to ASTM as class F or low calcium fly ash [23].
Loss on ignition (LOI) value unable measuring the unburned carbon in fly ash. Conversely, LOI value overestimates the amount of unburned carbon. This is due to some other reactions as dehydration of lime, decomposition of carbonates and oxidation of iron minerals through heating process. Therefore, more accurate ways as thermogravimetric analysis is nominated to replace LOI, which gives accurate data about the weight changes with temperature and time [24].

Table 1. Chemical composition of the fly ash.

Main constituents	Amount (Wt. %)
SiO_2	49.59
TiO_2	1.469
$Al_2 O_3$	14.14
Fe_2O_3	15.79
MgO	2.296
CaO	9.506
Na_2O	1.882
K_2O	1.515
P_2O_5	0.459
V_2O_5	0.642
Other	2.711

3.2 DTA and TGA Analysis

Figure 1a illustrates the TGA curves in which the onset degradation temperature of PS is 157 °C that increased by blending it with LDPE to be 192 °C. Table 2 shows that the onset degradation temperature of PS/LDPE 80/20 blend (C0) increased by embedding it with 10 wt.% (C10) and 50 wt.% (C50) of fly ash to be 197 °C and 210 °C, respectively. The increase in values of the onset degradation temperature for the blend (C0) and its composites of C10 and C50 more than PS is due to the reduction in the mobility of polymer chains by adding LDPE and fly ash, respectively. The formation of free radicals at weak bonds (usually peroxide groups and/or chain ends) is usually resulted from the thermal decomposition of polystyrene and LDPE. Consequently, this is followed by chain transfer reactions of the primary free radicals that transmit until the entire matrix is influenced. Embedding fly ash in the blend leads to the motion restriction of the chain segments which prevents the chain transfer reactions. Therefore, the thermal stability of the composites improved [25]. Hence, the degradation process is affected by the embedded fly ash which delays the degradation. This is owing to increase the thermal parameter of the composites with respect to the pristine blend itself, as seen in Table 2.

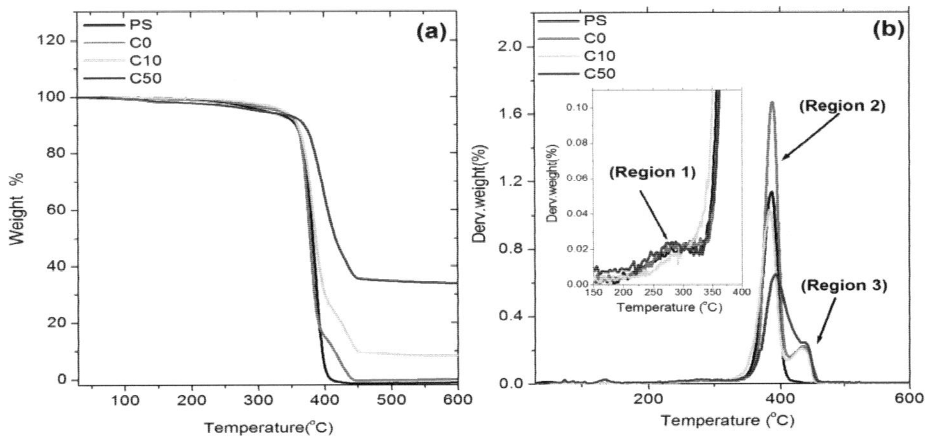

Figure 1. TGA curves **(a)** and DrTGA curves **(b)** for PS, PS/LDPE 80/20 blend (C0) and fly ash/PS/LDPE 10/80/20 composite (C10) and fly ash/PS/LDPE 50/80/20 composite (C50). The inset is the magnification of DrTGA curves from 150 to 400 °C.

Figure 1b shows clearly the thermal decomposition behavior of the samples obtained from the derivative thermogravimetric (DrTGA) curves. For PS, the weight loss started slowly from 157 °C to 321 °C (region 1) followed by a rapid degradation process up to 454 °C (region 2). PS has only one sharp peak, which means that the weight loss of PS takes place in a one-step degradation process from 321 °C to 454 °C, as observed in figure1b. This was confirmed by the observed values of the weight loss in Table 1 which showed the low value in the 1st region and the high value in the 2nd region. On the other hand, there are sharp and broad (or shoulder) peaks depending on the second and the third decomposition regions for the blend and its composites, as seen in figure 1b. Table 2 illustrates that the positions of the peaks in figure 1b rely on thermal degradation of LDPE, PS and fly ash. A slow degradation (region 1) was observed below 345 °C for the blend and its composites. Afterwards, a sharp degradation (region 2) was observed in temperature ranges of (318 - 407°C), (345 - 405°C) and (345 - 394°C) for the blend, the composites of C10 and C50, respectively. Then, broad peaks (region 3) take place for the blend and the composites of C10 and C50 at (407 – 478 °C), (405 - 548 °C) and (394-600 °C), respectively. The weight loss at the 2nd region was higher than that at the 3rd region for the blend and its composites as clearly observed in Table 2. As mentioned in a published literature [26], that the weight loss of LDPE was happened in a range of temperatures between 400 °C and 500 °C. Therefore, an overlap of degradation is expected to be existed between the 2nd region of both of PS and LDPE. This was responsible about the observed sharp peaks of the blend and its composites, as seen in figure 1b. For all the examined samples, the observed degradation in region 1 can be a result of the elimination of the absorbed moisture [27]. Table 2 shows that there was no decomposition residue for PS and its blend while it was about 9% and 33% for C10 and 50 C50, respectively. Therefore, the sample of the more thermal stability is the composite that contains the highest amount of fly ash (50 wt.%, C50).

Table 2. The degradation temperature, weight loss (%), total residue (%), the temperature of mass loss at 10% ($T_{10\%}$), 50% ($T_{50\%}$) and 65% ($T_{65\%}$), and maximum degradation temperature from DrTGA curves (T_{max}).

Sample (wt.%)	**Degradation temperature** (°C)	**Weight loss** (%)	**Total residue** (%)	$T_{10\%}$ (°C)	$T_{50\%}$ (°C)	$T_{65\%}$ (°C)	T_{max} **from DrTGA** (°C)
Ps	**Region 1: 157**-321 **Region 2:** 321-**454**	**Region 1:** 5 **Region 2:** 95	0	357	382	387	387
PS/LDPE 80/20	**Region 1: 192**-318 **Region 2:** 318-407 **Region 3:**407- **478**	**Region 1,2:** 84 **Region 3:** 16	0	369	388	393	388 434
Fly ash/ PS/LDPE 10/80/20	**Region 1: 197**-340 **Region 2:** 340-405 **Region 3:** 405- **548**	**Region 1,2:** 73 **Region 3:** 18	9	358	385	393	383 433
Fly ash/ PS/LDPE 50/80/20	**Region 1: 210**-340 **Region 2:** 340-394 **Region 2 :**394- **600**	**Region 1,2:** 32 **Region 3:** 35	33	367	416	477	392 439

3.3 Dielectric Properties

The polarity as being polar or non-polar has a great effect on dielectric properties of polymers. Polystyrene (PS) and Low density polyethylene (LDPE) are considered as non-polar polymers [7, 28]. The variation ε' with frequency at different constant temperatures for PS and C10, as an example for all the composites, are depicted in figures 2a & 2b, respectively. Regarding to these figures, the permittivity illustrated a discrepancy in relation to the high frequency and a similar characteristic at low frequency. At a lower frequency, < 50 kHz, the values of ε', for PS, its blend and its composites, was considerably high. This can be attributed to the asymmetry at phenyl side group of PS, thus it has a small dipole moment (0.26D), despite of being a non polar polymer [29, 30]. The high values of ε' at low frequencies can be interpreted on base of the interfacial polarization i.e., the building up of bound charges at interfaces within the material of the samples. This interfacial polarization occurs in heterogeneous structures due to the accumulation of charges at the interfaces between variant regions which are not similar to each other in its conductivity. Besides, at the interfaces between the sample and the electrodes, there is an accumulation of bound charges which is included in the net polarization (electrode polarization). An extra capacitive component is produced into the system by this space charge accumulation [31]. Conversely, for all the applied constant temperatures, the values of ε', decreased with increasing the applied frequency, >50 kHz, for all the composites, see figure 2b. This can be due to the dipole polarization which requires a considerable long time as the side groups of the main chains of the polymers lose slowly their capability to track the applied field. Consequently, their contribution to the polarization dies down and this resulted in decreasing the dielectric constant of the composites [31, 32]. However, the values ε' became independent of frequency for PS and its blend, at higher frequencies, >50 kHz, due to non-polarity structure of PS and LDPE. Also, figures 2a & 2b showed that, for PS and its composites, ε' decreased by increasing the applied temperature due to the low polarity of PS and LDPE [33].

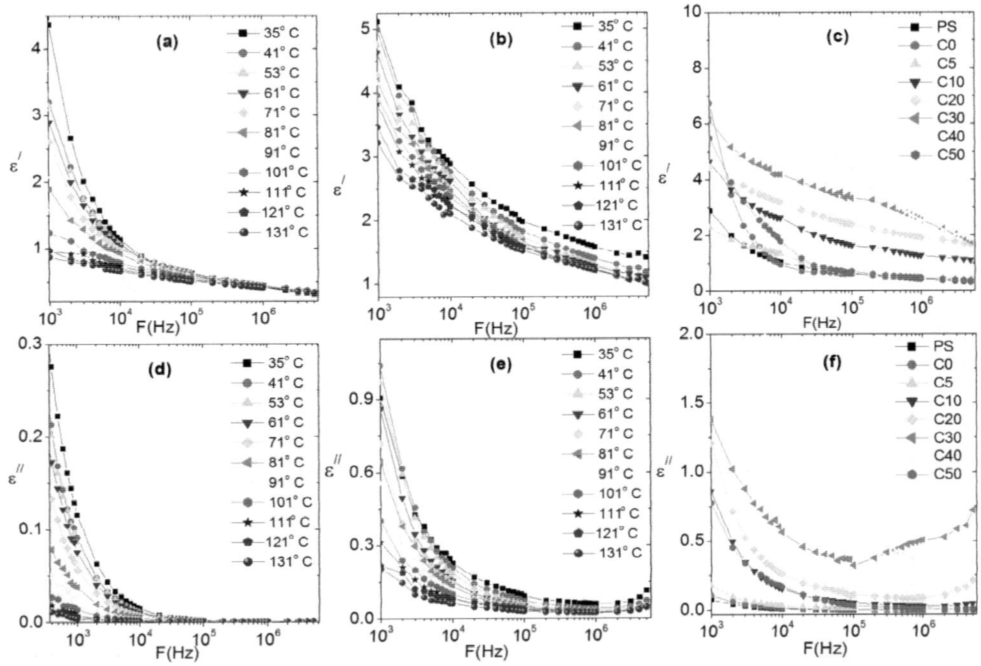

Figure 2. The variation of dielectric constant (ε') and dielectric loss (ε'') with frequency for PS **(a and d, respectively)** and fly ash/PS/LDPE 10/80/20 composite **(b and e, respectively)**. The variation of dielectric constant (ε') and dielectric loss (ε'') with frequency**,** for all the samples, at constant temperature of 61°C **(c and f, respectively)**.

The composition of the material as well as the reliance on the metal electrode type can obviously affect the dielectric properties of the polymers. For instance, the imaginary part (ε'') of the relative complex permittivity ε^*, depends on two important factors. The first one is the electrode resistance which can be minimized through using electrode material with high conductivity. However, the second one is the properties of the polymer itself. These two factors show frequency and temperature dependence [34]. The variation of ε'' with frequency, at variant constant temperatures, for PS and C10, as an example for all the composites, are presented in figures 2d & 2e, respectively. For PS and its composites, ε'' values were decreasing from 1 kHz up to 50 kHz followed by a stability until 5 MHZ for PS and up to 1 MHz for the composites. This can be attributed to the slower polarization mechanism as a result of stretching and twisting of the long carbon chains attached to the main polymeric chain [35]. However, a slight increase in the values of ε" were observed at frequencies > 1 MHz for all the composites. This can be a result of increasing the relaxation loss of the dielectric materials in case of applying a high frequency. In addition, at higher frequencies, ionic polarization can take place and consequently this encourages the charge carriers to instantly cross the surface of the composite. This can lead to an additional improving in the values of ε" at higher frequencies [36]. Figures 2c & 2f showed the variation of ε' and ε", respectively, with frequency for PS, the blend (C0) and the composites (C5 - C50), at 61°C. By increasing the content of fly ash in the blend the values of ε' and ε" increased up to the composite of C40 then they decreased for the higher fly ash composite (C50). The observed increase in ε' and consequently ε" can be due to the higher values of the dielectric constant of the metal oxides of the fly ash as SiO_2 (dielectric constant ≈ 3.9) and Al_2O_3 (dielectric constant ≈ 9) [17, 37].

The ln - ln plot of ε'' versus frequency, at different constant temperatures, are illustrated in figures 3a, 3b & 3c, for PS, C20, as an example for the composites, and all the samples at 61°C, respectively. Figures 3a & 3b shows a linear variation of plotting ln ε" versus ln ω. Thus, the frequency dependence of the imaginary part of the complex permittivity ε" can be presented by the law of fractional power as the following [38]:

$$\varepsilon'' = B\omega^{n} \quad (1)$$

Where B is a constant and n is a frequency power factor. The relation between ln ε and ln ω is known as Maxwell – Wagner formula for the heterogeneous media which is obtained by re-arranging Eq. 1 as the following [39]:

$$\ln \varepsilon'' = lnB + n\ ln\omega \quad (2)$$

$$n = -\frac{4\pi kT}{W_m} \quad (3)$$

Where W_M is the maximum barrier height which is defined as the energy required to move the electron from a site to the infinity. The values of n were determined from the negative slope of the straight lines obtained from ln – ln plot of ε'' versus frequency, see figures 3a, 3b & 3c. Table 3 shows that the deduced values of n and W_M were obviously temperature dependent as they were increasing by increasing the applied temperature. Also, Table 3 exhibits the effect of the fly ash content on the values of n and W_M for the composites which were all higher than those of the blend (C0) itself. Dominating polarization in a material by mobile hopping charge carriers may not display Debye type loss peak in their dielectric curves. Instead, a monotonic rise of both dielectric constant and loss values with decreasing frequency take place and hence it follows the fractional power law of ω^n. This attitude was firstly known by Jonscher [40-42] which is called the universal dielectric response.

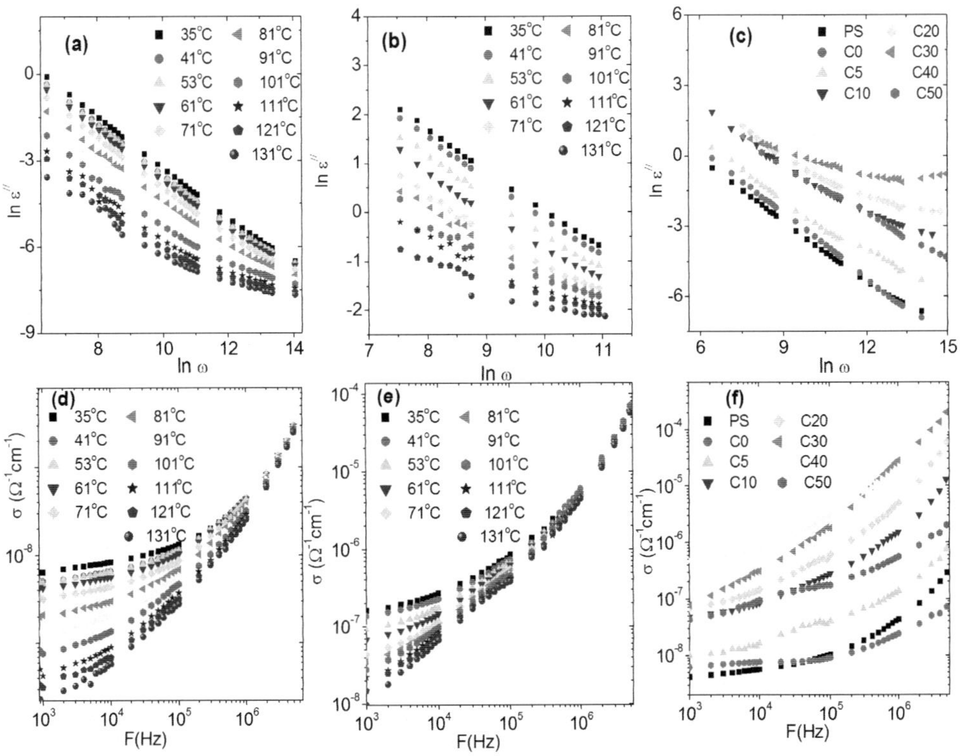

Figure 3. The variation of ln ε'' with ln ω and ac conductivity (σ_{ac}) with frequency for PS **(a and d, respectively)** and fly ash/PS/LDPE 20/80/20 composite **(b and e, respectively)**. The variation of ln ε'' with ln ω and ac conductivity (σ_{ac}) with frequency for all the samples at constant temperature of 61°C **(c and f, respectively)**.

3.4 AC conductivity

The values of ac conductivity (σ_{ac}) were obtained from the measured capacitance and the calculated dielectric loss [43]. The variation of ac conductivity, with a wide range of frequencies, from 1 kHz to 5 MHz, was examined at different temperature from 35 to 131° C. Figures 3d, 3e and 3f illustrate the frequency dependence of σ_{ac} for PS, C20 composite, as an example for all the composites, and all the different samples at 61°C, respectively. For each constant temperature, the ac conductivity exhibits a linear increasing in its value with increasing the applied frequency as a common behavior for all the samples. This behavior can be due to the presence of charge carriers that supply to the conduction mechanism along the polymer chains [7]. In addition, for all the samples, figures 3d & 3e showed that σ_{ac} was decreasing by increasing the applied temperatures. This behavior was due to the fact that the hopping distance increases with the thermal expansion of the polymer and polymer blend matrices. However, the more effect of the applied temperature on σ_{ac} was clearly displayed at low frequencies. However, σ_{ac} was nearly temperature independent for higher frequencies that existed above 1 MHz, for all the samples.

Table 3. The values of the exponent (n) and the barrier potential (W_m), at different temperatures, for PS, PS/LDPE blend and variant composites of fly ash/ PS/LDPE.

T°C	Parameter	PS	C0	C5	C10	C20	C30	C40	C50
41	n	0.158	0.041	0.154	0.226	0.172	0.591	0.149	0.309
	W_m	***0.192***	***0.169***	***0.191***	***0.209***	***0.196***	***0.396***	***0.190***	***0.234***
53	n	0.162	0.055	0.172	0.208	0.196	0.654	0.143	0.239
	W_m	***0.201***	***0.178***	***0.203***	***0.212***	***0.209***	***0.487***	***0.196***	***0.221***
61	n	0.170	0.083	0.266	0.223	0.236	0.603	0.142	0.325
	W_m	***0.208***	***0.188***	***0.235***	***0.222***	***0.226***	***0.434***	***0.201***	***0.255***
71	n	0.176	0.141	0.263	0.193	0.301	0.643	0.165	0.358
	W_m	***0.215***	***0.207***	***0.241***	***0.220***	***0.254***	***0.498***	***0.213***	***0.277***
81	n	0.23	0.216	0.320	0.193	0.323	0.581	0.146	0.403
	W_m	***0.237***	***0.233***	***0.269***	***0.226***	***0.270***	***0.436***	***0.214***	***0.306***
91	n	0.255	0.266	0.351	0.245	0.361	0.580	0.170	0.551
	W_m	***0.252***	***0.256***	***0.289***	***0.249***	***0.294***	***0.448***	***0.226***	***0.419***
101	n	0.232	0.343	0.372	0.298	0.390	0.670	0.128	0.570
	W_m	***0.251***	***0.294***	***0.307***	***0.275***	***0.317***	***0.585***	***0.221***	***0.449***
111	n	0.217	0.406	0.390	0.281	0.338	0.630	0.138	0.686
	W_m	***0.253***	***0.334***	***0.325***	***0.276***	***0.299***	***0.536***	***0.230***	***0.632***
121	n	0.185	0.471	0.558	0.352	0.621	0.535	0.24	0.687
	W_m	***0.249***	***0.385***	***0.460***	***0.314***	***0.537***	***0.438***	***0.268***	***0.650***

Also, at constant temperature of 61°C, figure 3f shows that increasing the fly ash content in the blend can improve σ_{ac} up to the composite C40 then it decreased. This can be attributed to the fact that increasing the fly ash content enables forming excess of charge carriers. So, the hopping conduction mechanism is dominant up to the composite of C40. However, for higher content of fly ash as C50, the filler particles may become large enough to be as obstacles for the hopping of charge carriers between the sites [44].

3.5 Complex Electrical Modulus Properties

Interfacial polarization plays an important role in the polymeric materials due to the additives, fillers or even impurities that make these systems to be heterogeneous. Generally, in some systems, interfacial relaxation is obscured by conductivity and the high dielectric permittivity at low frequencies. To overcome such a problem in studying the interfacial polarization, it has been decided to use the formalism "electric modulus" to study the polymer conductivity relaxation behavior [21]. In fact, for numerous cases, the dynamics of interest are better to be understood through plotting data in one of some choices, which are counterpart representations. The data illustration in M* (the electric modulus which is equal to 1/ε*) is physically comparable with the data depiction in ε* formalism. Nevertheless, in some cases, the data represented in M* becomes more helpful as the effects of the space charge can be restrained. Therefore, the conductivity current relaxation can be studied well [31, 45]. The complex electric modulus can be obtained according to the deduced formula by McCrum *et al.* [46] as the following:

$$M^* = \frac{1}{\varepsilon^*} = M^{/} + iM^{//} = \frac{\varepsilon^{/}}{\varepsilon^{/2} + \varepsilon^{//2}} + i\frac{\varepsilon^{//}}{\varepsilon^{/2} + \varepsilon^{//2}} \qquad (4)$$

Where $M^{/}$ and $M^{//}$ are the real and imaginary parts of the electric modulus, respectively. Owing to being the complex permittivity is in the denominator of Eq 4, its effect in dominating $M^{/}$ and $M^{//}$ has effectively lessened. Consequently, this permits more broad analysis of the dielectric data [21]. Herein, the complex modulus analysis relied on plotting $M^{/}$ and $M^{//}$ over a wide range of frequencies and different constant temperatures. The frequency dependence of $M^{/}$ and $M^{//}$ for PS, the composite C10, as an example for all the composites, and all the different samples at 61°C, are presented in figures 4a & 4d, figures 4b & 4e and figures 4c & 4f, respectively.

For all the samples, the $M^{/}$ values do not equal to zero at low frequencies and high temperature. Consequently, it is expected that the electrode polarization may take place for the all samples [47], see figures 4a, 4b & 4c. It can be noted that $M^{/}$ is increasing with increasing the applied frequency and with increasing the temperature for PS, its blend and composites. The $M^{//}$ frequency dependence for PS and its blend (C0), as seen in figures 4d & 4f, shows the existence of relaxation peaks at low frequency. However, the values of $M^{//}$ for the composite C10 were high at low frequency. Afterwards, they decreased up to ≈ 500 kHz followed by an obvious increasing at high frequencies without appearing any peak of relaxation, see figure 4e. This behavior is common for all the composites except for the composites of C5 and C50 which have a peak of $M^{//}$ at ≈ 59 kHz and 40 kHz, respectively. For all the observed relaxation peaks, the low-frequency side of the peak describes the region of frequencies where the hopping process of the charge carriers can occur from one site to the neighboring one. However, the charge carriers are confined to their potential wells and they can make localized motion within the potential well at the high frequency side of $M^{//}$ peak. The range of frequencies, where the peak itself is appeared, can explain the transition from long-range to short range mobility of the charge carriers [48, 49]. Figures 4c & 4f prove that the frequency dependence of $M^{/}$ and $M^{//}$, respectively, for all the samples, are depending on the filler content.

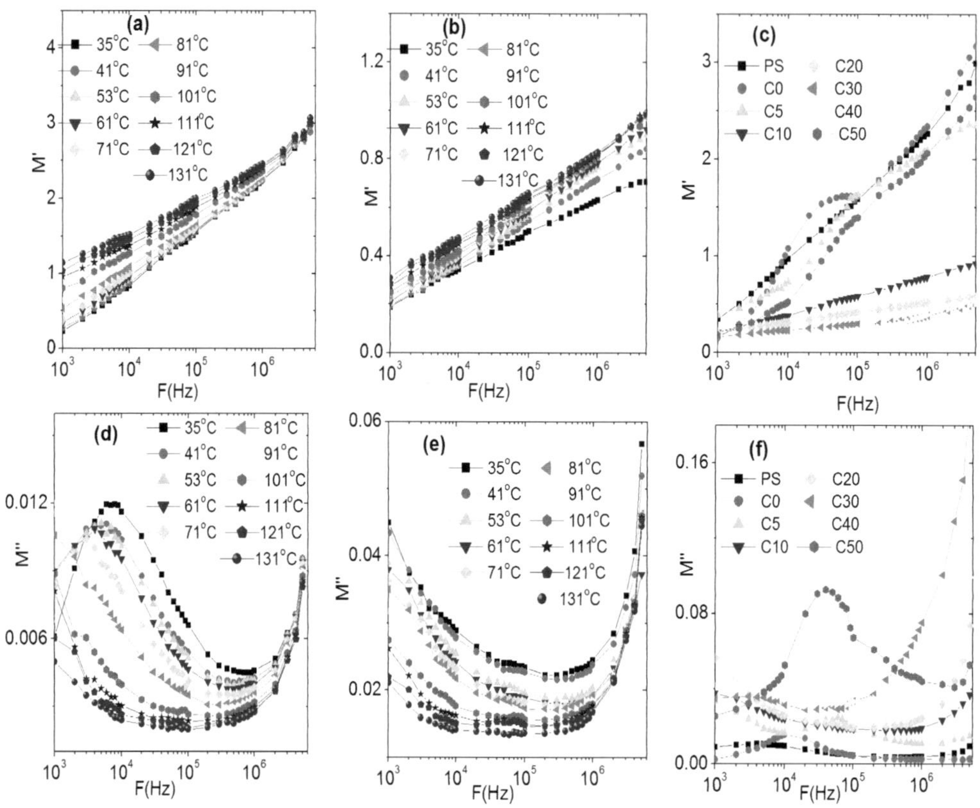

Figure 4. The variation of real (M') and imaginary (M'') parts of electric modulus for PS **(a and d, respectively)** and fly ash/PS/LDPE 10/80/20 composite **(b and e, respectively)**. The variation of real (M') and imaginary (M'') parts of electric modulus for all the samples at constant temperature of 61°C **(c and f, respectively)**.

4- Conclusion

The simple melt mixing method was used for preparing PS, PS/LDPE blend and composites of fly ash embedded in the blend with different contents. Analyzing the obtained data of XRF for the used fly ash showed that it can be considered according to ASTM as class F fly ash. TGA analysis illustrated that the degradation of the composites is affected by the filler content which delays the degradation process. This is due to increasing the thermal parameter of the composites with respect to the pristine blend itself. DrTGA curves showed that PS has only one sharp peak, which means that the weight loss of PS takes place in a one-step degradation process. However, there were sharp and broad peaks depending on the different decomposition regions for the blend and its composites. The values of ε' became independent of frequency for PS and its blend, at higher frequencies, >50 kHz. This can be attributed to the non-polarity structure of PS and LDPE. By increasing the content of fly ash in the blend, the values of ε' and ε'' increased up to the composite of C40 then they decreased for C50 composite. The frequency dependence of $M^{/}$ and $M^{//}$, respectively, for all the samples, are depending on the embedded content of the filler in the PS/LDPE blend.

References

[1]- Mirfakhrai M, Madden J D W, Baughman R H 2007 *J Mater. Today* **10** 30-38.
[2]- Amin M, Nasr G M, M. Sobhy S 1991 *J. Mater. Sci.* **26** 4615- 4620.
[3]- Gwaily S E, Nasr G M, Badawy M M 2001 *Egypt J Sol* **24** 193-205.
[4]- Koralege R S H and Jayasuriya C K 2015 *Journal of Emerging Trends in Engineering and Applied Sciences* **6** 248- 252.
[5]- Ho Ba T, Roberts T K, Lucas S 2018 *Crit Rev Biotechnol.* **38** 308–320.
[6]- Sharma J, Chand N, Bapat M N 2012 *Results in Physics* **2** 26–33.
[7]- Ahmed R M, Morsi R M M 2017 *Mod. Phys. Lett. B* **31** 1750278.
[8]- Rozenberg B A, Tenne R 2008 *Prog. Polym. Sci.* **33** 40.
[9]- Fleming M S, Mandal T K, Walt D R 2001 *Chem. Mater.* **13** 2210.
[10]- Elashmawi I S, Hakeem N A, Marei L K, Hanna F F 2010 *Physica B* **405** 4163.
[11]- Fan B, Liu Y, He D, Bai J 2017 *Materials* **10** 838.
[12]- Folorunso O, Hamam Y, Sadiku R, Ray S S, Joseph A G 2019 *Polymers*, **11**, 1250.
[13]- Sengupta S, Pal K, Ray D, Mukhopadhyay A 2011 *Compos. B. Eng.* **42** 1834–1839.
[14]- Chauhan S R, Kumar A, Singh I, Kumar P 2010 *J Minerals and Materials Characterization and Engineering* **9** 365-387.
[15]- Deepthi M V, Sharma M, Sailaja R R N, Anantha P, Sampathkumaran P, Seetharam*u S* 2010 *J. Materials and Design* **31** 2051–2060.
[16]- Rohatgi P K, Matsunaga T, Gupta N 2009 *J. Materials Science* **44** 1485–1493.
[17]- Bachtrong P, Nguyenthanh L, Phamthi H 2017 *Chemical Engineering Transactions* **56** 1207-1212.
[18]- Singh A P, Kumar A, Chandra A, Dhawan S K 2011 *AIP Advances* **1** 022147.
[19]- Sanjib B, Madhushree K, Dubey T. 2011 *Indian J. Phys.* **85** 559-573.
[20]- Ahmed R M 2012 *Int. J. Mod. Phys. B* **26** 1250159.
[21]- Nasr G M, Ahmed R M 2010 *Mod. Phys. Lett. B* **24** 911–919.
[22]- Tanikella P, Olek J 2017 *Updating physical and chemical characteristics of fly ash for use in concrete* (Joint Transportation Research Program Publication No. FHWA/IN/JTRP-2017/11). West Lafayette, IN: Purdue University.
[23]- Alehyen S, EL Achouri M, Taibi M 2017 *Journal of Materials and Environmental Sciences* **8** 1783-1796.
[24]- Straka P, Náhunkova J, Žaloudkova M 2014 *Thermochimica Acta* **575** 188-194.
[25]- Koralege R S H, Jayasuriya C K 2015 *Journal of Emerging Trends in Engineering and Applied Sciences* **6** 248- 252.
[26]- Salmah H, Lim B Y, The P L 2013 *J Thermoplast Compos. Mater.* **26** 1155–1167.

[27]- Giannakas A, Salmas C, Leontiou A, Tsimogiannis D, Oreopoulou A, Braouhli J 2019 *Nanomaterials* **9** 1105.
[28]- Novák I, Krupa I, Luyt A S 2005 *J. Appl. Polym. Sci* **95** 1164–1168.
[29]- Bur A J 1985 *Polymer* **26** 963.
[30]- Work R N, Trehu Y M 1956 *J. Appl. Phys.* **27** 1003.
[31]- Saafan S A, El-Nimr M K, El-Ghazzawy E H 2006 *J. Appl. Polym. Sci.* **99** 3370–3379.
[32]- Khare P K, Jain S K 2000 *Bull. Mater. Sci.* **23** 17–21.
[33]- Shujah-Aldeen B A 2007 *Study the Effect of Temperature and Frequency on the Dielectric Properties of Some Commercial Polymers* (M.Sc. Thesis: University of Sulaimani Iraq).
[34]- El Sayed A M, Morsi W M 2013 *Polym. Compos.* **2013** 2031-2039.
[35]- Liu J, Zheng H, Zhang Y, Wei H, Liao R. 2017 *Energies* **10** 1526.
[36]- Singha S, Thomas M J 2008 *IEEE Trans. Dielectr. Electr. Insul.* **15** 2–11.
[37]- Nawaz M 2015 *Active and Passive Electronic Components* **2015** 651527.
[38]- Ganaie M, Zulfequar M 2015 *Acta Phys. Pol. A* **128** 59-63.
[39]- Baraker B M, Lobo B 2019 *Bull. Mater. Sci.* **42** 18.
[40]- Jonscher A K 1999 *J. Phys. D: Appl. Phys* **32** 57–70.
[41]- Jonscher A K 1983 *Dielectric Relaxation in Solids* (Chelsea Dielectrics: London UK).
[42]- Rizvi T Z, Abdul Shakoor 2009 *J. Phys. D: Appl. Phys.* **42** 095415.
[43]- Zidan H M, Ghamaz N A, Abdel Ghany A M, Lotfy A 2014 *International Journal of Physics and Research* **4** 7-14.
[44]- Murugendrappa M V, Khasim S, Prasad M V N A 2005 *Bull. Mater. Sci.* **28** 565–569.
[45]- Neagu R M, Neagu E 2000 *J Appl Phys* **88** 6669.
[46]- McCrum N G, Read B E, Williams G 1967 *An Elastic and Dielectric Effect in Polymeric Solid* (Dover: New York USA).
[47]- Hanafy T A 2012 *J. Appl. Phys* **112** 034102.
[48]- Tsangaris G M, Kouloumbi N, Kyvelidis S 1996 *Mater. Chem. Phys.* **44** 245.
[49]- Bouaamlat H, Hadi N, Belghiti N, Sadki H, Bennani M N, Abdi F, Lamcharfi T, Bouachrine M, Abarkan M 2020 *Adv Mater Sci Eng* **2020** 8689150.

IOP Conf. Series: Materials Science and Engineering **956** (2020) 012003 doi:10.1088/1757-899X/956/1/012003

Characterization and Performance of PAni-TiO_2 Photovoltaic Cells Treated by RF Plasma

F. M. El-Hossary [1], Ahmed Ghitas [2], A. M. Abd El-Rahman [3,1], A. A. Ebnalwaled [4,5], M. Abdelhamid Shahat [2,*]

[1] Physics Department, Faculty of Science, Sohag University, Sohag, Egypt.
[2] PV Unit, Solar and Space Research Department, National Research Institute of Astronomy and Geophysics (NRIAG), Helwan, Cairo, Egypt.
[3] King AbdulAziz University, Jeddah, KSA.
[4] Electronics & Nano Devices Lab, Physics Department, Faculty of Science, South Valley University, Qena, 83523 Egypt.
[5] Egypt Nanotechnology Center (EGNC)/ Cairo University Sheikh Zayed Campus, 12588 Giza, Egypt.

* E-mail: m.abdelhamid999@gmail.com

Abstract. Severe research attempts are still in progress to improve the performance of polyaniline (PAni) based photoactive layers as one of the cheapest materials used for manufacturing organic solar cells. Herein, polymer solar cells were fabricated with ITO/(PAni-TiO_2)/Au system. The photoactive layers (PAni-TiO_2) were treated with a hydrogen-plasma discharge for low processing time of 0, 3 and 5 min to enhance the synthesized solar cells efficiency. The morphology, microstructure and optical properties of the prepared samples and plasma treated nanocomposite layers were investigated and discussed. The performance of bulk heterojunction (BHJ) cell samples have been systematically investigated before and after plasma treatment. The absorption and optical band gap energy is increased after the treated PAni-TiO_2 photoactive layers. It is found that, the efficiency was enhanced to 0.7% after 5 min of hydrogen plasma process compared to 0.36% for the pristine cell. The efficiency increase is ascribed to a structural change that accompanied by a rapid increase in surface roughness, which leaded to a decrease in the reflected photons and in turn an increase in the produced charge carriers.

Keywords: TiO_2 nanoparticles (NPs), Polyaniline (PAni), Nanocomposites, Hydrogen plasma, BHJ solar cells.

1. Introduction

Over the last decades, great research efforts have been exerted to synthesize more efficient conducting polymers for fabricating organic photovoltaic (OPV) solar cells. Polyaniline (PAni) is a promising type of conducting polymers used for potential semiconductor applications owing to its high conductivity, simple preparation, good environmental stability [1]–[4].

Recently, inorganic nanoparticles (NPs) were successfully introduced into polymer based-materials in order to construct an efficient photoactive nanocomposite layer for improving organic photovoltaic performance [5], [6]. Titanium dioxide (TiO_2) NPs with nanocrystalline structure have found to be efficient semiconductor material for fabricating organic solar cells for their unusual chemical and physical properties and surface nanofeatures [7]–[9]. The desired properties of TiO_2 can be easily modified to meet different application needs through optimizing the preparation process conditions that control NPs size, shape and their distribution [10]. Based on the extraordinary properties of both PAni and TiO_2 NPs, the PAni-TiO_2 nanocomposites have potential applications in charge storage, conductive coating, photovoltaic solar cell and electrocatalytic applications [11], [12]. In addition, semiconductor NPs or nanocomposite thin films consisting of NPs have been treated using RF plasma discharge to improve the photoactive properties of these materials to accomplish special performance needs [13]. For example, several conductive metal oxide thin films have been exposed to hydrogen plasma treatment in order to increase density carrier and oxygen vacancies; leading to an increase in their corresponding carrier mobility [14], [15]. Further, low-pressure plasma discharge has been used with N_2/SO_2 gas mixture to modify the TiO_2 NPs for providing them with sulfonation and to enhance their dispersion in PAni matrix [16].

The present work focuses on using low-pressure low-temperature RF plasma processing to treat PAni-TiO_2 nanocomposite films using hydrogen gas. Preliminary attempts for the improvement of photooptical properties are carried out with applying hydrogen plasma treatment for short processing time. Structural, surface morphology, electrical and optical properties of the as prepared and treated nanocomposite PAni-TiO_2 layers are measured and discussed to evaluate the photoactive layer performance. Moreover, J-V measurements of PAni-TiO_2 layers are obtained in dark and under simulated sunlight at a total light intensity of 100 $mW.cm^{-2}$ to evaluate the effect of hydrogen plasma treatment on the solar cell efficiency.

2. Materials, Preparation and Characterization Techniques

2.1. Materials

Titanium dioxide powder (titanium powder, Ranbaxy > 5μm) was used as precursor for preparation of TiO_2 NPs. Chemical materials of aniline monomer (C_6H_7N, Purity 99+ %), hydrochloric acid HCl (37%) and ammonium peroxydisulfate (APS) with a chemical formula of $(NH_4)_2S_2O_8$ were purchased from Sigma Aldrich, Germany.

2.2. Preparation of TiO_2 NPs

A solution of TiO_2 was prepared by adding a 10.0 g TiO_2 powder in 40 ml distilled water in a glass vial under magnetic stirring for a reaction time of 1h at room temperature as shown in Fig. 1. In order to ensure good dissolution, the mixture was allowed to sonicate for several minutes. Then, it is thermally heat treated at 75 °C for 4 hours in a Teflon lined stainless steel autoclave. Afterward, the autoclave was naturally allowed to cool down to room temperature for 24 h. The precipitation was filtered out, washed several times with ethanol and distilled water, and then calcinated at 500 °C for 3 h. The previous procedure for producing TiO_2 NPs was repeated two times (8 and 12 h) to study the reaction time effect on the photovoltaic properties of the PAni-TiO_2 nanomaterials.

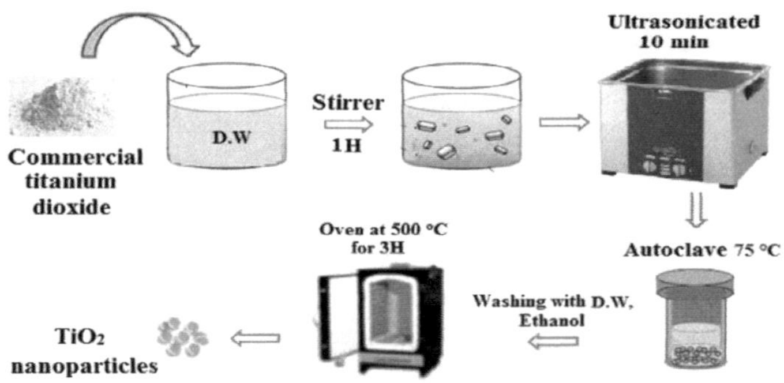

Figure 1. Synthesize of TiO_2 NPs using the hydrothermal method

2.3. Synthesis of PAni and PAni-TiO_2 Nanocomposite

Chemical oxidative polymerization method was used to prepare pure PAni and PAni-TiO_2 nanocomposites, where aniline monomer was polymerized by acidic catalyst of HCl in the presence of ammonium persulfate (APS) as a strong oxidizing agent. APS is a colorless inorganic salt and it is a highly soluble material in water. A 0.01 mol of aniline monomer is dissolved in a 0.01 mol of HCl and mixed using an ultrasonic unit (Elmasonic S 40H) at room temperature. A 0.01 mol of APS was dissolved in a 15-mL distilled water to prepare a dilute APS solution. Finally, in order to obtain the pure PAni solution, a dilute APS solution was placed into a burette and added drop wise to the mixture of aniline and HCl with a slow drop rate, under magnetic stirring in an ice bath at a reaction time of 5.5 h. The mixing process between the reactive mixtures is obtained with a slow drop rate and continued under stirring in order to ensure good oxidization and homogenous polymerization of PAni solution, respectively. To prepare PAni-TiO_2 nanocomposites, TiO_2 powder was added to the pure PAni solution. Then, the obtained composites were deposited by a spin coating on ITO coated conducting glass substrates by using an 800 rpm for 30s. Finally, the obtained films were dried at 50 °C. A 0,01 mol aniline monomer is mixed in a 0,01 mol HCl solution using an ultrasonic unit (Elmasonic S 40H) at room temperature. Within another beaker, A 0.01 mol of APS was dissolved to create a diluted solution of APS within 15 mL distilled water. TiO_2 powder was spread in the beaker to cover the TiO_2 NPs with conductive PAni as a shell, which produce aniline monomer and was put on a magnetic stirrer for 30 min. In order to confirm complete, the interaction between the aniline and TiO_2 NPs, TiO_2 has been blended into the aniline blends in the connection of 20% vs. 80% aniline solution, followed by several minutes in ultrasound. A dilute APS solution was put in a burette and was applied dropwise to the aniline – HCl – TiO_2 mixture under magnetic stirring (1000 rpm) at a reaction time of 5.5 h in an ice bath. Accordingly, PAni-TiO_2 nanocomposites can be produced using a typical physicochemical technique [17], that starts with a slow drop rate of about 1 drop / second and continues to be stirred to ensure good oxidation and homogeneous polymerization of PAni solution. At the end of the physicochemical reaction the color is observed changing to dark green. This means that an oxidation phase is underway that ends with the PAni-TiO_2 composite solution for the formation process. The precipitate composites were filtered and washed to extract excess acid with distilled water and ethanol. A spin coating on an ITO coated conducting glass substrata with 800 rpm for 30 s was deposited for the last composite solutions, followed by a drying process with 50 °C for 30 min. The methods of deposition and drying were performed in the air atmosphere.

2.4. Plasma surface modification for PAni-TiO_2 nanocomposite films

The photoactive layer consisting of the PAni-TiO_2 nanocomposite was deposited with a thickness of 150 nm and directly introduced into a plasma reactor chamber. Then, the chamber was evacuated to a base pressure of 20×10^{-3} mbar and filled with a reactive gas of H_2 at a working pressure of 82×10^{-3} mbar. The hydrogen plasma discharge was generated with an inductively coupled technique using a rf power generator (model HFS 2500 D, 13.65 MHz). Different plasma processing times of 0, 3 and 5 min are applied on the as-prepared films for a fixed processing power 220 W. The treatment temperature was measured to be 423 K; using a Chromel-Alumel thermocouple locate in contact with the sample surface. At the end of the plasma process, the treated samples were allowed to cool down under evacuation for few minutes. Finally, in order to construct an entire solar cell system, the coated ITO (80-82% transmittance and 4×10^{-5} Ω.cm resistivity, sigma Aldrich) with photoactive layers has to be covered by a rectangular shaped Au thin film. The conductive material Au coat film was used as a highly effective cathode and deposited by a thermal evaporation method. A schematic of the stack device architecture is shown in Fig. 2.

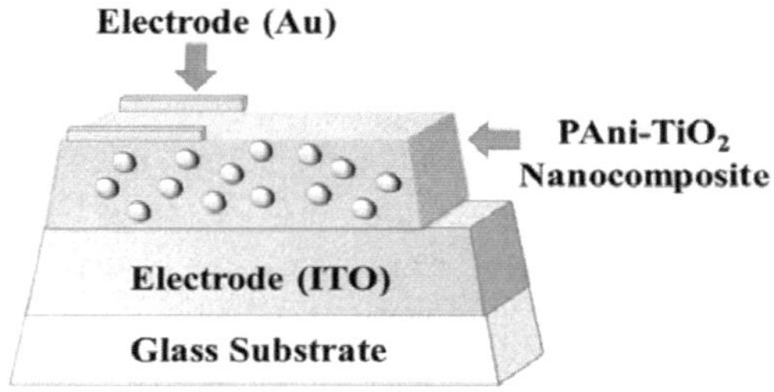

Figure 2. Organic solar cell architecture of ITO/PAni-TiO_2/Au.

2.5. Characterization

The phase and structure of the samples were examined by X-ray diffraction (XRD), using Philips diffractometer with a Cu target (λ=1.54056 Å) and a graphite monochromator at 40 kV and 30 mA. The scanning speed and step are 2.5°/min and 0.06°. The functional groups of inorganic NPs, pristine and treated nanocomposite thin films are evaluated using Fourier Transform Infrared (FT-IR) Spectroscopy (Jasco Model 4100_Japan). The FT-IR analysis is obtained from 4000 to 400 cm^{-1} with a resolution of 4 cm^{-1} at room temperature. Further, the surface morphology has been characterized by high resolution scanning electron microscope (SEM) (JOUL SEM model JSM-5500 - Japan), operated at an accelerated voltage of 10 KV. The optical properties of the as prepared and treated nanocomposite films were measured by a computerized SPECORD 200 PLUS spectrophotometer at room temperature with a 1 nm step and at a normal incidence angle in the wavelength range 190 – 1100 nm. The Profilometer (Talysurf 50-Taylor Hopson precision) is carried out as a mechanical test machine, and used to extract the topographical changes in the sample surface. In this study, the roughness parameter, Ra, represents the average roughness of the nanocomposite films. Furthermore, the J –V characteristics of the solar cell under a fixed illumination intensity of 1.0 Sun at 25°C is measured using a solar simulator system. The simulation test is carried out using a xenon lamp as an ideal light source for solar spectrum with the use of a KEITHLEY 2400 Source Meter.

3. Results and discussion

3.1. Microstructural analysis

Fig. 3 shows the typical XRD patterns of PAni-TiO_2 nanocomposite films treated by hydrogen plasma at processing time of 3 and 5 min compared to that of the pristine sample. Diffraction PAni peaks with different intensities and peak widths are detected in all samples between 10° and 80° owing to parallel and perpendicular periodicity of the polymer chain [18]. These diffraction peaks are found to be correspond to (110), (012), (003), (102) and (323) planes. After surface treatment, changes in the diffraction peaks are observed and found to be related to TiO_2 and corresponding to (002), (102), (103), (201) and (004) planes. The crystallite size (D) of TiO_2 NPs can be calculated using the Debye–Scherrer equation [19]:

$$\boldsymbol{D} = \frac{0.94\,\lambda}{\beta\,cos(\theta)} \qquad (1)$$

Where **λ** is the wavelength, **β** is the full width at half maximum of the peak, and θ is the diffraction angle. The degree of crystallinity and crystallite size of TiO_2 NPs achieved 86.6% and 11.86 nm, respectively.

The polycrystalline diffractions match well with the standard powder diffractions of tetragonal TiO_2 phase with anatase crystal phase (a = b = 3.7335 Å, c = 9.3788 Å, α = β = γ = 90.0) (ICSD, Code no 9852) [20], and orthorhombic polyaniline [21]. The diffraction peak at 2θ of 23°is due to polymer chains placed parallel and closed to each other producing crystalline zones surrounded by amorphous areas [22]. The polycrystalline diffractions of the pristine and

IOP Conf. Series: Materials Science and Engineering **956** (2020) 012003 doi:10.1088/1757-899X/956/1/012003

treated PAni-TiO_2 nanocomposites are identical to those caused by the standard powder diffractions of tetragonal anatase TiO_2 structure and orthorhombic PAni. This indicates that no phase transformation involved in the hydrogen plasma surface treatment. Generally, the crystallinity in a particular sample relies on the periodicity, which implies a long-range sequence of atoms, ions or molecules. As noted, the XRD pattern of the plasma-treated nanocomposite films showed an increase in the crystallization and peaks intensity with increasing plasma processing times. The strong effects on the surface of PAni-TiO_2 of hydrogen plasma leading to changes in the chemical bonds in the nanocomposite films between PAni and TiO_2 and the physical dependence of the polymer on the area of TiO_2 NPs are attributed to this fact.

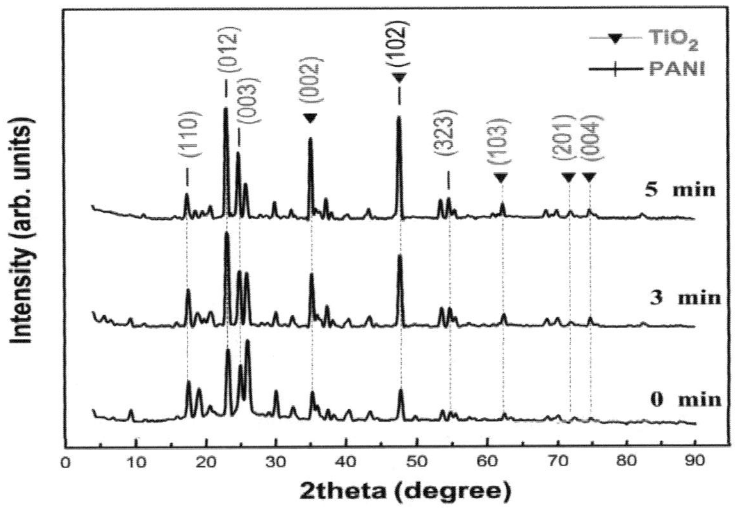

Figure 3. XRD patterns of plasma treated and pristine PAni-TiO_2 nanocomposite samples

3.2. FT-IR spectroscopy

The composition of the pure PAni, pristine nanocomposite and plasma treated PAni-TiO_2 nanocomposite films is characterized by FT-IR spectroscopy and the obtained spectra is shown in Fig. 4. FT-IR spectra was obtained in the range 4000-400 cm^{-1} in order to confirm the composition of the chemical compounds that formed before and after plasma treatment. The FT-IR spectra show that, the main peaks of the pure PAni domains continued to appear clearly in the pristine and treated PAni-TiO_2 nanocomposites with hydrogen plasma.

FT-IR spectra of pure PAni exhibits the main absorption peaks at the region 3121–472 cm^{-1} and continued to distinctly appear in the pristine and treated nanocomposites with intensity changes and a small peak shift towards higher wavenumbers, which assigned to the presence of TiO_2-NPs in the nanocomposite structure. The wide absorption peak between 3690–2983 cm^{-1} can be attributed to the stretching O–H and the hydrogen-bond N–H stretching of aromatic amine [23], [24]. Compared to the pristine sample, the peak at 3121 cm^{-1} is detected with a wide stretching shape and a small shifting towards short wavenumber. The peak broadening and shifting effect is higher for sample treated at processing time of 5 min with that of 3 min. This might be attributed to the attraction by a group of adjacent atoms or two nearby functional groups [25]. On the other hand, this peak shifting has found to be associated with the highest crystallinity of the nanocomposite compound that found in Fig. 3, which in turn leads to an increase in the energy gap and results an increase in the light absorption. Photo-induced holes and hydroxylic radicals may be produced on surfaces of hydroxyl bonds [26]. It allows these radicals to improve the composite's photocatalytic efficiency [27]. The characteristic peaks at 1537 cm^{-1} and 1412 cm^{-1} are assigned to C=N and C=C ring asymmetric and symmetric stretching vibrations, due to the presence of quinonoid structures and benzenoid ring, respectively. Whereas the peaks at

1301 cm^{-1} and 1193 cm^{-1} correspond to C–N stretching of the benzenoid and quinonoid unit of doped PAni, respectively. This calculated the delocalization rates and characterized as a PAni conductivity peak [28]. In the spectrum of PAni-TiO_2 nanocomposite, the appearance of PAni main peaks (i.e. C=C and C–N peaks) confirmed PAni formation on nanoTiO_2 surface [29]. The good interactions between PAni and TiO_2 NPs resulted in a coordination between the N atoms of PAni chains and the TiO_2 NPs shielding the nanoparticles with a PAni layer coating [30], [31]. The peaks at 860 cm^{-1} is associated with C–C and C–H in the benzenoid unit [17]. Whereas, the peak at 616 and 472 cm^{-1} was relatively changed compared to pure PAni, which may be attributed to the out-of-plane bending flexural vibration of C–H aromatic rings of PAni and vibrations in the stretching Ti–O–Ti bonds of TiO_2 [23]. Thus, indicates a structural change in the polymer matrix resulting from a strong H-bonding interaction exists between PAni and TiO_2-NPs.

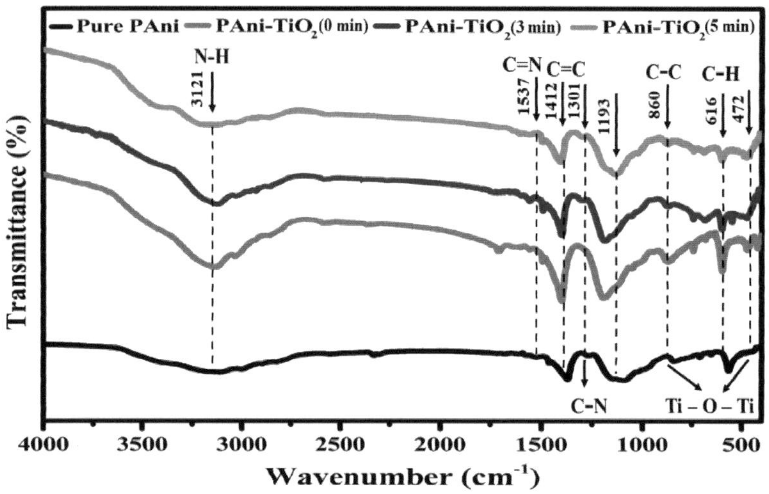

Figure 4. FT-IR spectra of pure PAni and nanocomposites films were treated by H-plasma for different processing times

3.3. *Morphology*

Fig. 5 illustrates SEM images for the pure PAni film, pristine and plasma treated PAni-TiO_2 nanocomposite thin films. Based on previous literature, an increased active layer with a micro-structure results in a relatively rough surface, which increases the surface contact between the layers of the nanocomposite and the electrode surface [32]. In the microporous surface of the active layer of the composite, the increased capacitance can take place as a result of maximum absorption [33]. The morphology of the nanocomposite particles depends on the conditions of polymerization such as pH, temperature, polymerization time as well as the TiO_2 structure and ratio [34]. Besides, the white area is assigned to be TiO_2, while the grey and black areas refer to the polymer chain. As observed, the surface morphology of PAni-TiO_2 shows a featured structure with non-agglomerated TiO_2 NPs and some porous fibrous networks. These results are consistent with previous results indicated that with the presence of a PAni shell on the surface of TiO_2 NPs, which causes repulsion between the nanoparticles to prevent their agglomeration [33], [35], [36]. However, after hydrogen plasma treatment, the surface morphology of the samples is drastically changed and the needle structure is appeared. Further, the structure of PAni-TiO_2 nanocomposite changes from an agglomerated needle to mostly needle structure with the variation of plasma processing time from 3 min to 5 min; offering suitable plasma energy to the imbedded nanoparticles for the formation of needle structure.

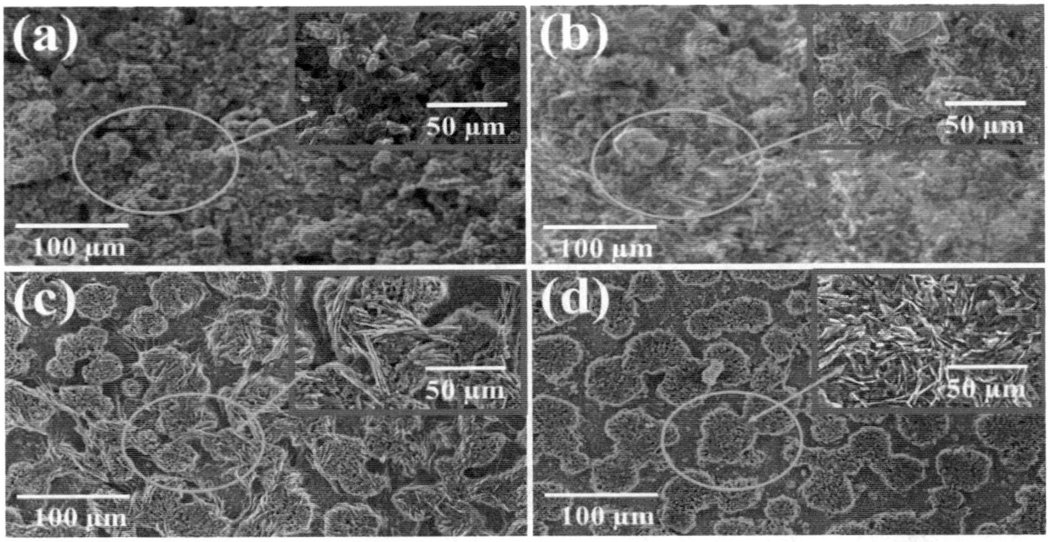

Figure 5. SEM images of (a) Pure PAni film, (b) Pristine and (c, d) Treated PAni-TiO_2 nanocomposite samples at 3- and 5-min using hydrogen plasma surface treatment, respectively.

3.4. Roughness measurements

Fig. 6 shows the effect of hydrogen plasma treatment on the roughness factor (Ra) of PAni-TiO_2 nanocomposite films as a function of plasma treatment time. Ra is abruptly increased with the increase of plasma treatment time. As it is found in other works, the change in surface roughness is mostly ascribed to the structural and morphological changes [37]. The increase in surface roughness of PAni-TiO_2 nanocomposites that composed of elastic/plastic hybridized materials is expected to be a physical response to heat effects produced by a RF-electric field generated plasma. This is mostly acceptable in case of using a low melting point elastic material like polymer combined with plastic NPs of metal oxides. The flow of heat stress in polymer matrix is mostly a reason for a fast-mechanical response and in turn partially responsible for increased surface roughness. The high heat capacity of TiO_2 (0.69 kJ/kg K) [38], can be considered as another contributing factor for the high increase in R_a of plasma treated PAni-TiO_2 nanocomposites. The interpretation of this dissimilar trend is mostly ascribed to the difference in the structural features of both nanocomposites. The agglomeration of TiO_2 NPs in PAni matrix and their size effect leads to an increase in surface roughness [39]. However, the fibrous network structure is mostly responsible for surface smoothness. The change in the surface roughness of the nanocomposite films affects their optical properties including transmittance, absorbance and optical constants [40], [41].

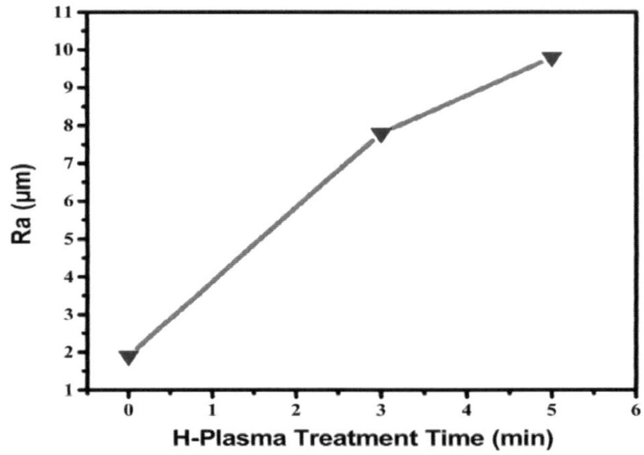

Figure 6: Roughness factor (R_a) of plasma treated and pristine PAni-TiO_2 nanocomposite samples

3.5. Optical Properties

3.5.1. Optical constants

Fig. 7 demonstrates the absorbance spectra obtained for pristine and hydrogen plasma treated nanocomposite samples for different wavelengths. The pristine and treated nanocomposite films reveal similar absorbance behavior with distinct absorbance values. Accordingly, their transmittance values are expected to have different values as well. The absorption and transmittance values as a function of the wavelengths of the different plasma processing time show a change proportional to those of the pristine nanocomposite film of PAni-TiO_2 as one can see in Fig.7a, b. As shown in from Fig. 7a, a broader absorption peaks are appeared with a similar spectrum in the range of 300–800 nm. The nanocomposite films exhibit wide absorption peaks at 320 nm and 595 nm with limited peak shifts towards lower wavelengths with increased hydrogen plasma surface treatment up to 5 min compared to the pristine nanocomposite film. This peak shift indicates the effect of the plasma modification on the electronic structure of TiO_2 NPs in the valence band that could interact with the higher occupied molecular orbital (HOMO) of PAni [42]. Moreover, the nanocomposites show equivalent energy absorption peaks at wavelengths of 320, 595 and 740 nm which in turn leads to transitions of $\pi\rightarrow\pi^*$ within the benzenoid rings, Polaron$\rightarrow\pi^*$ of quinoid rings, and $\pi\rightarrow$polaron excitonic of the quinoid rings, respectively [11]. Determination of optical energy gap and the transition type were obtained using Tauc equation [43]:

$$\alpha h\nu = B\left(h\nu - E_g\right)^{1/2} \qquad (2)$$

Where **hv** is photon energy and ***B*** is a transition probability dependent parameter, α is the absorption coefficient, $\mathbf{E_g}$ is the optical band gap [37].

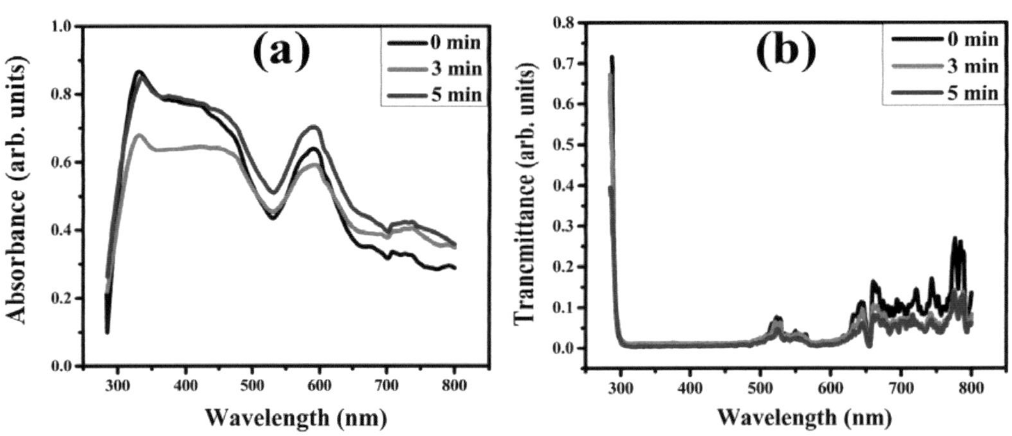

Figure 7. (a) The absorbance and (b) transmittance behaviors obtained for plasma treated and pristine nanocomposite samples

Fig. 8 shows a variation in the optical band gap of pristine and treated PAni-TiO_2 films with hydrogen plasma at 3 and 5 min. The optical band gap (E_g) is estimated from the strong absorption region, which depends on the short-range order in the amorphous state and defect state [44]. It is found to be increased with the increase of plasma treatment time and has values of 1.99, 2 and 2.16 eV at treatment time of 0, 3 and 5 min, respectively. These results may be attributed to the reduction of the system disorder and the increase of the defect state density. As a result of a relatively wide band gap, the PAni-TiO_2 nanocomposites films have large visible transmittance [45].

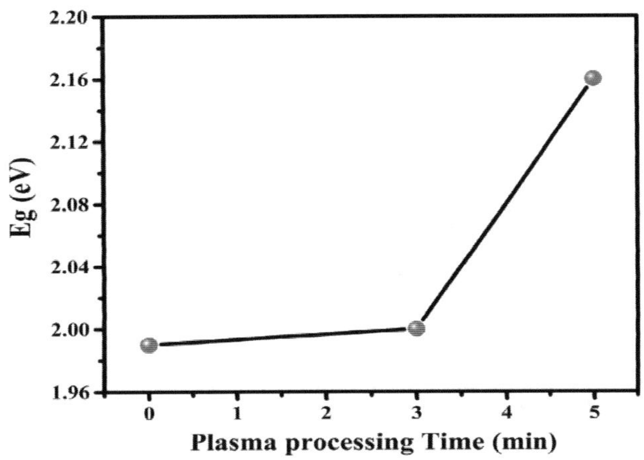

Figure 8. Optical band gap of pristine and treated PAni-TiO_2 nanocomposite by hydrogen plasma for processing time of 3 and 5 min

3.5.2. Optical dispersion characteristics

The refractive index (***n***) and the extinction coefficient (***k***) are considered as the most efficient optical dispersion parameters for many applications of optoelectronics. For polymer-based materials, the ***n*** has a direct effect on physical, chemical, and molecular characterization and in turn on their optical, electrical, and magnetic properties. The optical constants ***(n, k)*** are estimated from reflectance and absorption coefficient based on the famed equation which developed by Wemple and DiDomenico [46].

$$\left(n^2(hv) - 1\right)^{-1} = \frac{-1}{E_0 E_d}(hv)^2 + \frac{E_0}{E_d} \qquad (3)$$

Where ***h*** is the Plank's constant, and ***hv*** is the photon energy, E_0 is the average excitation energy for electronic transitions or the oscillator energy, E_d is the dispersion energy which measures the average strength of inter-band optical transitions or the oscillator strength, which is a measure of the intensity of the inter band optical transitions, does not depend significantly on the band gap. The dispersion curves of ***n*** and ***k*** versus hv for hydrogen plasma PAni-TiO_2 samples for different treatment times are demonstrated in Fig. 9. As observed, ***n*** and ***k*** parameters of the PAni-TiO_2 nanocomposites are highly affected after the hydrogen plasma process compared to the untreated one. These results may be attributed to changes in the absorption coefficient as a result of the plasma process which are directly proportional to the optical dispersion parameters ***(n, k)*** of the PAni-TiO_2 films [47].

IOP Conf. Series: Materials Science and Engineering **956** (2020) 012003 doi:10.1088/1757-899X/956/1/012003

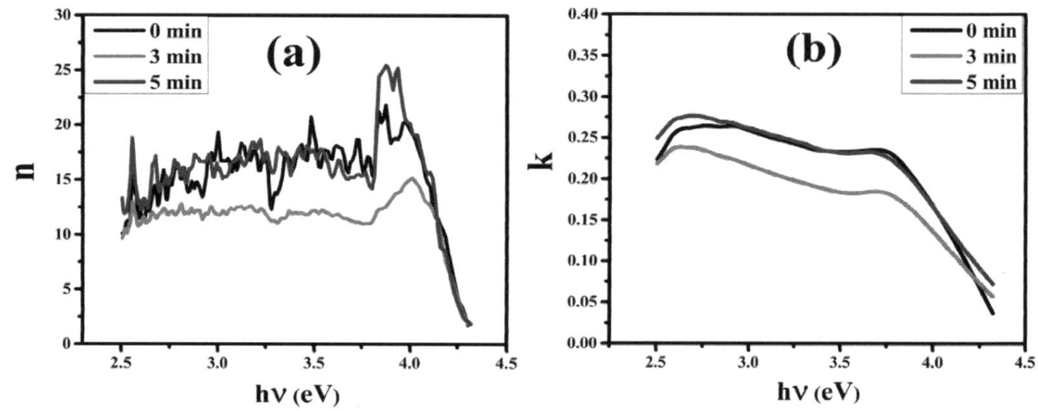

Figure 9. (a), (b) Dispersion curves of n and k versus photon energy ($h\upsilon$), respectively, for PAni-TiO_2 samples treated using hydrogen plasma for processing time of 3 and 5 min.

3.5.3. Optical conductivity

Optical conductivity σ_{opt} is a complex optical quantity that significant for describing optical properties of materials and their applications. The complex optical conductivity is composed of real and imaginary parts ($\sigma_{opt} = \sigma_1 + i\sigma_2$); both are positive and dielectric constant dependent (ε) [48]. The real (σ_1) and imaginary (σ_2) parts can be expressed as in the following equations [48]:

$$\sigma_1 = \omega\varepsilon_2\varepsilon_0 \ , \ \sigma_1 = \omega\varepsilon_1\varepsilon_0 \qquad (4)$$

Where ω is the angular frequency. As seen from the previous equations, σ_1 and σ_2 are photon energy dependent parameters. Fig. 10 expresses σ_1 and σ_2 as a function of photon energy ($h\upsilon$). In general, one can see that both real and imaginary parts have appeared with fewer fluctuations after hydrogen plasma treatment compared to that of the pristine nanocomposite film. Further, they have a maximum value appears in a broad peak located close to the end of the applied photon energy region. Then, their values are abruptly decreased. Furthermore, the plot has a plateau region free from any fluctuations or with little fluctuations is detected for treated samples compared to that of the pristine film. With more precise description, the behavior of the pristine sample is nearly comparable to that of the treated sample at treatment time of 5 min. Further, the lowest values of σ_1 and σ_2 optical conductivity are found for the PAni-TiO_2 treated at treatment time of 3 min.

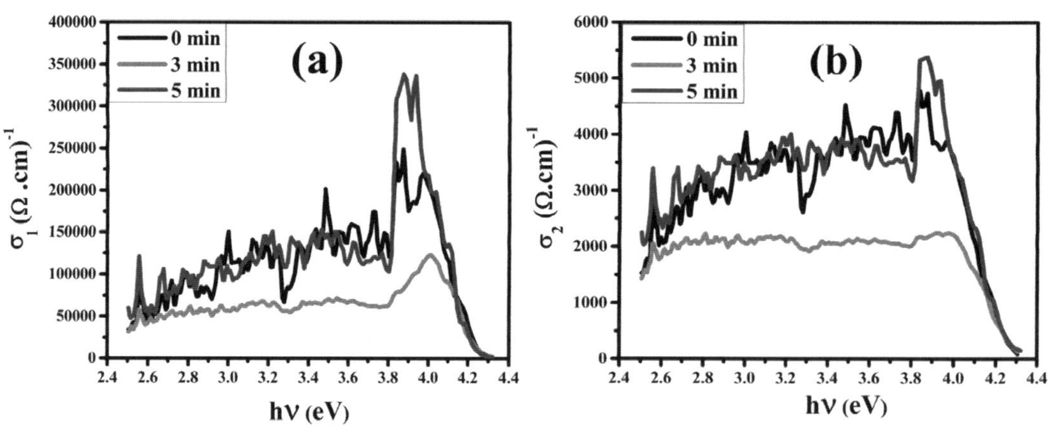

Figure 10. (a) Real and (b) imaginary optical conductivity of pristine and plasma treated nanocomposite films versus photon energy ($\boldsymbol{h\upsilon}$)

The optical conductivity ($\boldsymbol{\sigma_{opt}}$) and electrical conductivity ($\boldsymbol{\sigma_{elc}}$) can be determined according to the following equations [49], [50]:

$$(\sigma_{opt}) = \frac{\alpha n c}{4\pi} \quad , \quad (\sigma_{elc}) = \frac{2\lambda\sigma_{opt}}{\alpha} \qquad (5)$$

Fig. 11 shows the value of the optical and electrical conductivity of the treated nanocomposite films as a function of the photon energy. As found, the behavior of optical and electrical conductivity is photon energy dependent too. The electrical conductivity increases over a low energy range of visible light (until 2.7 eV), then the $\boldsymbol{\sigma_{elc}}$ starts to gradually decrease with the increase of visible light energy. This behavior continues within the relatively low energy-UV until the energy of 3.7 eV, followed by a sudden decrease with further increase of photon energy. As known, the solar light reaching the earth surface combines about 40% of visible and less than 10% from the ultraviolet radiation. Therefore, the visible light has substantial effects on optical properties of the active layer of nanocomposite metal oxides. The increase in conductivity as a function of visible photon energy has to be ascribed to the enhancement in the transition process of the charge carriers by absorbing the energy of the incident photons. Accordingly, the excess conductivity can be called optical conductivity. The optical conductivity increases with the increase in photon energy within visible region.

This behavior continues until reaching a maximum value at about 3.7 eV and then it is abruptly decreased. After hydrogen plasma treatment, the optical conductivity is only improved in the visible region for the sample treated at treatment time of 5 min. This is positively reflected on the light absorption in the active layer and certainly on the solar cell efficiency. The improvement in the optical properties of PAni nanocomposites films leads to extend their photonic applications [51], [52].

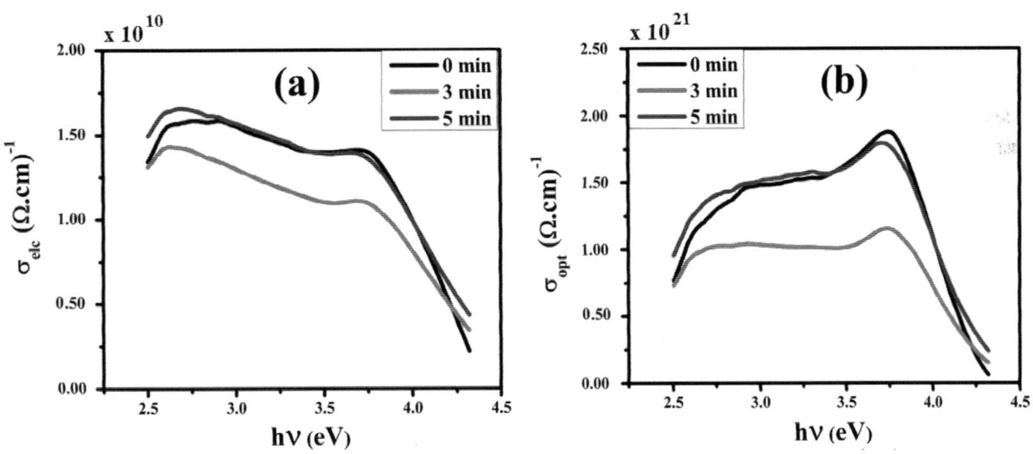

Figure 11. (a) Optical and (b) electrical conductivity of pristine and treated PAni-TiO_2 nanocomposite films as a function of photon energy

IOP Conf. Series: Materials Science and Engineering **956** (2020) 012003 doi:10.1088/1757-899X/956/1/012003

3.6. Performance evaluation of PAni-TiO_2 nanocomposite photovoltaic cells

Current density-voltage (J-V) characterization is a common technique used to determine the performance of solar cells. J-V measurements of solar cells are obtained in dark and under simulated sunlight with a total light intensity of 100 mW/cm^2 and with a solar spectrum under Air Mass 1.5 global. The A M 1.5 G is the solar spectrum arriving on the Earth's surface after passing the atmosphere at a zenith angle of about 48° [53], [54]. Herein, PAni's hybrid photoactive layers were designed as a donor mixed with TiO_2 NPs as acceptor in the fabricated solar cells. Fig. 12 shows the current density-voltage (J-V) characteristics of the solar devices structured by pure PAni, PAni-TiO_2 nanocomposite layers in dark. As observed, it exhibits the same asymmetric diode-like behavior.

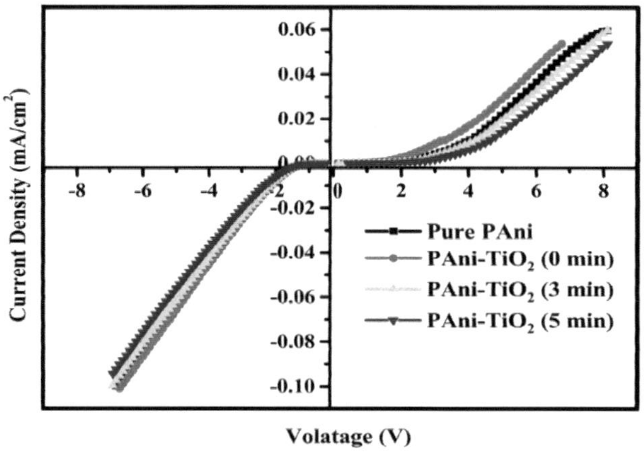

Figure 12. Diode-like behaviour of J-V curves obtained in the dark of designed photovoltaics based on pure PAni film, untreated nanocomposite film (pristine) and treated PAni-TiO_2 nanocomposite films using hydrogen RF-plasma at different processing times.

Fig. 13 shows the J–V curves obtained from photovoltaics based on pure PAni and pristine composite PAni-TiO_2 layers under illumination. As observed, the J–V curve is shifted towards higher values of current densities for the pristine composite device structured by ITO /(PAni-TiO_2) 0 min/Au compared to that one fabricated with ITO/Pure PAni /Au. These results show that the presence of TiO_2 NPs into the PAni matrix increases the carriers transport in PAni structure significantly through the contribution of polarons and bipolarons that move at a smaller distance in polymer chains [55]. Moreover, the J-V curves gradually shifted towards higher values of current densities of 4.19 and 4.71 mA/cm^2 for device structured by the photoactive layer treated with hydrogen plasma at 3 and 5 min, respectively, compared to 3.15 mA/cm^2 which obtained for a pristine layer designed cell. This shifts vertically under illumination due to light induced current generation. The possible reason of the higher current density may be ascribed to a reduction in the interfacial resistance between TiO_2 and PAni with the increase in the surface treatment time due to the increase in the number of polarons in PAni structure and the structural homogeneity of nanocomposites [2].

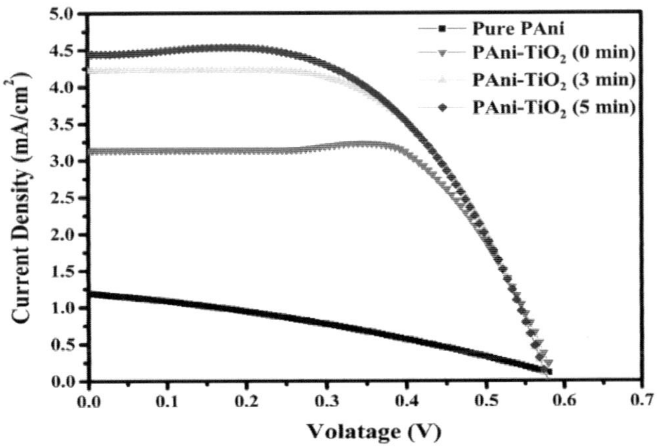

Figure 13. J-V characteristic curves of pure PAni film, pristine and treated PAni-TiO_2 solar cell devices obtained in the standard conditions and in a simulated light.

Photovoltaic parameters of the solar cell device such as short circuit current density (J_{SC}), open circuit voltage (V_{OC}), fill factor (FF) and power conversion efficiency (η) can be attained from J–V characterization curves. V_{OC} depends mostly on the trade-off energy between highest occupied molecular orbital (HOMO) of the donor and lowest unoccupied molecular orbital (LUMO) of the acceptor. Table (1) illustrates a comparison of photovoltaic parameters for the fabricated solar cells.

Herein, measurements of V_{OC} showed nearly constant values as the photoactive material (PAni) and have constant value (around 0.56 V) for all designed photovoltaic devices [56]. As observed, the photovoltaics efficiency of ITO/pure PAni/Au is 0.24% and improved to 0.36% by incorporating the pristine TiO_2 NPs into PAni structure. This is due to an increase of photocurrent density within composites structure [2]. The solar cells designed with PAni-TiO_2 nanocomposites treated for 3 min exhibited a gradual improvement in efficiency values of 0.49% and achieved the optimum value of 0.7% at a treatment time of 5 min. In contrast to other similar devices manufactured from active layers PAni-TiO_2 [57]–[59], these fabricated devices show better performance based on their measured efficiency. This performance improvement might be ascribed to the following reasons:

(i) The increase in the number of polarons and bipolarons in the PAni structure, which creates an efficient charge carries transfer network in the nanocomposite material. This interprets the conduction mechanism of the PAni-TiO_2 nanocomposite [60].

(ii) The rapid increase in surface roughness decreases the reflected photons and thus increases the produced charge carriers.

Table 1. Characteristic parameters for pure PAni, pristine and treated PAni-TiO_2 nanocomposites solar cells

Time (min)	V_{oc} (V)	J_{sc} (mA/cm^2)	FF (%)	η(%)
ITO/pure PAni/Au	0.56	1.20	35.1	0.24
ITO/(PAni-TiO_2) 0 min/Au	0.56	3.15	20.1	0.36
ITO/(PAni-TiO_2) 3 min/Au	0.56	4.19	21.0	0.49
ITO/(PAni-TiO_2) 5 min/Au	0.56	4.71	26.2	0.70

IOP Conf. Series: Materials Science and Engineering **956** (2020) 012003 doi:10.1088/1757-899X/956/1/012003

4. Conclusion

The effects of RF hydrogen plasma treatment on structural and optical properties of PAni-TiO_2 nanocomposites have been investigated. The as-prepared PAni--nanocomposite thin films were achieved with a mixing ratio of 20% TiO_2 at a reaction time of 12 h. After hydrogen plasma treatment, the crystallinity is improved with the increase of plasma treatment time. Further, morphology and roughness of the obtained nanocomposites is highly affected. The absorption and optical band gap have higher values for the treated PAni-TiO_2 photoactive layers compared to that of the pristine film. Consequently, the efficiency was enhanced to 0.7% after 5 min of hydrogen plasma process compared to 0.36% for the pristine cell. The improvement in photovoltaic efficiency is mostly ascribed to structural effects beside the rapid increase in surface roughness, which decreases the reflected photons and in turn increases the produced charge carriers. As a future objective, it is essential to optimize the structural and optical properties of the unmodified PAni-metal oxide nanocomposite thin films. In addition, intensive research work should be concentrated on optimizing the plasma treatment conditions of RF hydrogen discharge for the treatment of the photoactive PAni nanocomposite thin films.

References

[1] Mais L, Mascia M, Palmas S, Vacca A 2019 *Sep Purif Technol* **208** 153–159.
[2] Nemade K, Dudhe P, Tekade P 2018 *Solid State Sci* **83** 99–106.
[3] Aparna S, Elakhya N, Gopal G, Rajesh P, Ramasamy P 2018 *Optik (Stuttg)* **157** 1219–1226.
[4] Cui W, He J, Wang H, Hu J, Liu L, Liang Y 2018 *Appl Catal B Environ* **232** 232–245.
[5] El-Hossary F M, Ghitas A, Abd El-Rahman A M, Ebnalwaled A A, Fawey M H, Shahat M A 2020 *IOP Conference Series: Materials Science and Engineering* **762** 012001.
[6] Xu J, Wang J, Mitchell M, Mukherjee P, Jeffries-El M, Petrich JW, et al 2007 *J Am Chem Soc* **129** (42) 12828–12833.
[7] Haider A J, AL–Anbari R H, Kadhim G R, Salame CT 2017 *Energy Procedia* **119** 332–345
[8] Haider AJ, Thamir AD, Najim AA, Ali GA 2017 *Plasmonics* **12** (1) 105–115.
[9] Singh R, Dutta S 2018 *Fuel* **220** 607–620.
[10] Sundaram GA, Maniarsu S, Vijendar RP, Veerappan G, Karthikeyan V, Nomura K, et al 2018 *Catal Today* 0920–5861.
[11] Xia H, Wang Q 2002 *Chem Mater* **14** (5) 2158–2165.
[12] Deivanayaki S, Ponnuswamy V, Ashokan S, Jayamurugan P, Mariappan R 2013 *Mater Sci Semicond Process* **16** (2) 554–559.
[13] Makabe T, Petrovic ZL 2014 *CRC Press* **26**.
[14] Van de Walle CG 2000 *Phys Rev Lett* **85** (5) 1012.
[15] Singh AK, Janotti A, Scheffler M, Van de Walle CG 2008 *Phys Rev Lett* **101** (5) 55502.
[16] Hong YC, Lee SWSH, Kwon OP, Lee SWSH, Uhm HS 2010 *Surf Coatings Technol* **205** S271–S274.
[17] Yang C, Dong W, Cui G, Zhao Y, Shi X, Xia X, et al 2017 *Electrochim Acta 247* 486–495.
[18] Mostafaei A, Zolriasatein A 2012 *Prog Nat Sci Mater Int* **22** (4) 273–280.
[19] Buerger MJ, Azaroff LV 1958 *New York: McGraw-Hill* **238** 342.
[20] Hanawalt JD, Rinn HW, Frevel LK 1938 *Ind Eng Chem Anal Ed* **10** (9) 457–512.
[21] Selvan ST, Mani A, Athinarayanasamy K, Phani KLN, Pitchumani S 1995 *Mater Res Bull* **30** (6) 699–705.
[22] Basavaraja C, Pierson R, Venkataraman A, Basavaraja S 2009 *Macromol Res* **17** (8) 609–615.
[23] Shi L, Wang X, Lu L, Yang X, Wu X 2009 *Synth Met* **159** (23–24), 2525–2529.
[24] Kao S-Y, Lin Y-S, Chin K, Hu C-W, Leung M, Ho K-C 2014, *Sol energy Mater Sol cells* **125** 261–267.
[25] Socrates G 2004 *John Wiley & Sons.*
[26] Zhang Q, Bao N, Wang X, Hu X, Miao X, Chaker M, et al 2016 *Sci Rep 6* 38066.
[27] Seif A, Nikfarjam A, ghassem HH 2019 *Sensors Actuators B Chem* **298** 126906.

[28] Quillard S, Louarn G, Lefrant S, MacDiarmid AG 1994 *Phys Rev B* **50** (17) 12496.
[29] Monfared AH, Jamshidi M, Hashemi Monfared A, Jamshidi M, Monfared AH, Jamshidi M 2019 *Prog Org Coatings* **136** 105257.
[30] Zhang L, Liu P, Su Z 2006 *Polym Degrad Stab* **91** (9) 2213–2219.
[31] Min S, Wang F, Han Y 2007 *J Mater Sci* **42** (24) 9966–9972.
[32] Allen JE, Ray B, Khan MR, Yager KG, Alam MA, Black CT 2012 *Appl Phys Lett* **101** (6) 63105.
[33] Olad A, Behboudi S, Entezami AA 2012 *Bull Mater Sci* **35** (5) 801–809.
[34] Wang H, Han B, Lu J, Wu P, Cui W 2020 *Mater Lett* **260** 126906.
[35] Al-Ghamdi AA, Mahmoud WE, Yaghmour SJ, Al-Marzouki FM 2009 *J Alloys Compd* **486** (1–2) 9–13.
[36] Akimov YA, Koh WS 2010 *Nanotechnology* **21** (23) 235201.
[37] Fox M 2002 *American Association of Physics Teachers* 1269–1270.
[38] Bernard F, Delobelle P, Hirsinger L, Rousselot C 2008 *Materials Science Forum* **583** 213–228.
[39] Jlassi K, Radwan AB, Sadasivuni KK, Mrlik M, Abdullah AM, Chehimi MM, et al 2018 *Sci Rep* **8** (1) 1–13.
[40] Deshpande NG, Gudage YG, Sharma R, Vyas JC, Kim JB, Lee YP 2009 *Sensors Actuators B Chem* **138** (1) 76–84.
[41] Chieruzzi M, Cerritelli GF, Miliozzi A, Kenny JM 2013 *Nanoscale Res Lett* **8** (1) 448.
[42] Rahman KH, Kar AK 2020 *Mater Sci Semicond Process* **105** 104745.
[43] Pankove JI 1975 *Courier Corporation.*
[44] M. Reda S, M. Al-Ghannam S 2012 *Adv Mater Phys Chem* **02** (02) 75–8.
[45] Jamshidinia M, Kovacevic R 2015 *Surf Topogr Metrol Prop* **3** (1) 4003.
[46] Wemple SH, DiDomenico Jr M 1971 *Phys Rev B* **3** (4) 1338.
[47] Abdullah OG, Saber DR 2012 *Applied Mechanics and materials* **110** 177–182.
[48] Sarkar S, Das NS, Chattopadhyay KK 2014 *Solid state Sci* **33** 58–66.
[49] Sun Y-P, Riggs JE, Rollins HW, Guduru R 1999 *J Phys Chem B* **103** (1) 77–82.
[50] Kramer IJ, Sargent EH *Chem Rev* **114** (1) 863–882.
[51] Gupta K, Jana PC, Meikap AK 2010 *Synth Met* **160** (13–14) 1566–1573.
[52] Krishnakumar V, Nagalakshmi R 2005 *Spectrochim Acta Part A Mol Biomo. Spectrosc* **61** (3) 499–507.
[53] Ebnalwaled AA, Thabet A 2016 *Synth Met* **220** 374–383.
[54] Snaith HJ 2012 *Energy Environ Sci* **5** (4) 6513–6520.
[55] Diantoro M, Suprayogi T, Taufiq A, Fuad A, Mufti N 2019 *Mater Today Proc* **17** 1197–1209.
[56] Elumalai NK, Uddin A 2016 *Energy Environ Sci* **9** (2) 391–410.
[57] Geethalakshmi D, Muthukumarasamy N, Balasundaraprabhu R 2016 *Mater. Sci. Semicond. Process* **51** 71–80.
[58] Ibrahim M, Bassil M, Demirci UB, Khoury T, Moussa GEH, El Tahchi M, et al *Mater* 2012 *Chem Phys* **133** (2–3) 1040–1049.
[59] Liu Z, Zhou J, Xue H, Shen L, Zang H, Chen W 2006 *Synth Met* **156** (9–10) 721–723.
[60] Stojanović GM, Radetić MM, Šaponjić Z V, Radoičić MB, Radovanović MR, Popović Ž V, et al 2020 *Appl Sci* **10** (12) 4392.

IOP Conf. Series: Materials Science and Engineering **956** (2020) 012004 doi:10.1088/1757-899X/956/1/012004

Fano Resonance Properties in (K_3C_{60} / $HgBa_2Ca_2Cu_3O_{10}$) 1D photonic crystal

Arafa H Aly[1,*], D. Mohamed [1], N. S. Abd El-Gawaad[2], Z. S. Matar[3], Y. Trabelsi[4,5] and M. A.Mohaseb[1,3]

[1] TH-PPM group, Physics Department, Beni-Suef university, Beni Suef, Egypt
[2] Faculty of Science, King Khalid University, Abha, Saudi Arabia
[3] Umm-Al-Qura University, faculty of applied science, department of Physics, Mecca, Saud Arabia.
[4] College of Arts and Sciences in Muhail Asir, Physics Department, King Khalid University, Abha, Saudi Arabia.
[5] University of Tunis El Manar, National Engineering School of Tunis, Photovoltaic and Semiconductor Materials Laboratory, 1002, Tunis, Tunisia.
E-mail: arafa.hussien@science.bsu.edu.eg

Abstract

The Fano resonance and Electromagnetic induced reflectance (EIR) properties in one-dimensional superconductor photonic crystals (SPCs) have been investigated theoretically by Transfer Matrix Method (TMM). The periodic structure consists of alternating of pair superconductor materials are made of (K_3C_{60} / $HgBa_2Ca_2Cu_3O_{10}$) one-dimensional photonic crystal. The SPCs is enclosed by dielectric cap layer at different induced fields. To exam the efficiency of the reported structure, different parameters are used for analysis such as layers thicknesses, temperature, angle of incidence and the dielectric constant, dielectric cap layer on the EIR and Fano line shape. The investigation results exhibits tunable Fano resonances and EIR resonance peaks accompanied by asymmetrical line shape very sensitive to dielectric cap layer, constituent materials and dependence incidence angle. This will be useful reference for different applications of photonic topological states in the integrated photonic devices and information processing chips.

1. Introduction

Photonic crystals (PCs) are the objects of various theoretical and experimental researches due to their ability to control the flow of electromagnetic waves (EMWs). Such microstructures open a promising application in modern optics [1-9] such as optical filter [10, 11], laser LEDs [12], optical microcavities [13], modulators [14], and sensors [15,16].

In the past decades, the superconductor have introduced in photonic crystals due to their advantages and ability to reduce the damped EMWs instead of metal or normal materials. These new superconducting PCs provide an enlarged reflection bands can be tuned by the temperature of superconductor [17-22]. So, the response of system is mainly dependent on the London penetration depth, which is a function of the temperature and the external magnetic field as well [23-28]. Furthermore, the transmittance spectrum of considered superconducting PCs gives an asymmetric line-shape called Fano resonance and a quantum destructive interference phenomenon called electromagnetic induced reflectance (EIR) [29].

The Fano resonance appears in PCs result on interference between a continuum band of state and a discrete quantum state and shows a sharp change of intensity light [29-30]. These Fano resonances given by superconducting PCs are originated of interesting applications in optical switching, sensing and filtering [31, 32]. However, the tunable electromagnetically induced reflectance (EIR) is used specially for metamaterial and plasmonics [33].

In this paper, we predict and demonstrate the generated electromagnetic induced reflectance resonance (EIR) and Fano resonance of the stacked (K_3C_{60} / $HgBa_2Ca_2Cu_3O_{10}$/dielectric cap layer) on glass substrate. We have using two-fluid model and transfer matrix method to characterize and determine the photometric spectrum through the multilayered stacks.

2. Theoretical Models

In this part, we present the theoretical model of the proposed one dimensional superconductor PCs in the form of $(AB)^N$ terminated by dielectric cap layer. Here, layer A and layer B are set to be a superconductor with thickness $d_1$and d_2, respectively repeated for N periods. Figure (1) display the schematic diagram of the proposed periodic 1D superconducting PC structure consists of two superconductor materials. The first one is for (K_3C_{60}) and the second layer is from ($HgBa_2Ca_2Cu_3O_{10}$).

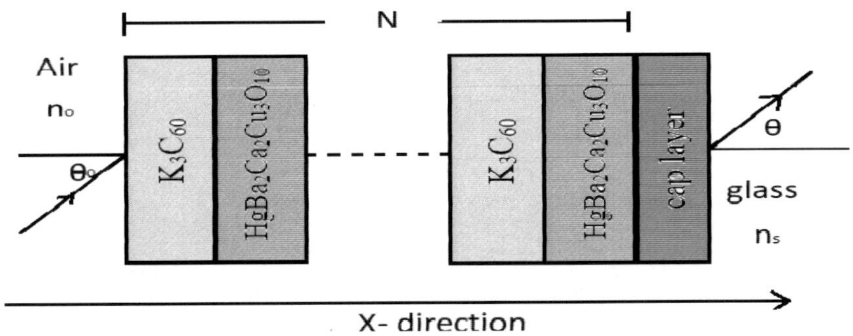

Figure 1: A Schematic diagram of the proposed 1D superconducting PC

The Gorter Casmir two fluid model is adopted to describes the electromagnetic response of the superconductor materials [18]. By adopting some approximations the relative permittivity of superconductor can be written as:

$$\varepsilon_s(\omega) = 1 - \frac{c^2}{\omega^2 \lambda_L^2} \quad (1)$$

Where, c and λ_L denote the velocity of light in vacuum and the temperature dependent London penetration depth [18] noted as follows:

$$\lambda_L = \frac{\lambda_0}{\sqrt{(1-(\frac{T}{T_c})^q)}} \quad (2)$$

With λ_0 represent the London penetration depth where $T = 0$ K. Here T and Tc are the operating and the critical temperature of the superconducting materials. With q takes 2 for High Tc superconductivity and 4 for low Tc superconductivity. The interaction between the incident EMWs and each interface layer through the periodic heterostructure is described using the following matrix [18]:

$$M_j = \begin{pmatrix} \cos(k_j\delta_j) & -(i/p_j)\sin(k_j\delta_j) \\ -(ip_j)\sin(k_j\delta_j) & \cos(k_j\delta_j) \end{pmatrix} \quad (3)$$

Where, M_j is the characterized matrix of the jth layer. With δ_j represents the phase variation at j^{th} layer . For both TE and TM modes, $\delta_j = d_j n_j \cos\theta_j$. Wheras, $p_j = n_j \cos\theta_j$ for TE mode and $p_j = \frac{\cos\theta_j}{n_j}$ for TM mode. θ_j is the angle of incidence.

The interaction between all stratified layers and incident EMWs is described using the following matrix [18]:

$$M = \begin{pmatrix} m_{11} & m_{12} \\ m_{21} & m_{22} \end{pmatrix} = (M_A M_B)^N \quad (4)$$

With N is the period of periodic superconducting layers. Thus, the transmittance values are given by the elements of transfer matrix as [18]:

$$t = \frac{2p_0}{(M_{11} + M_{12}p_f)p_0 + (M_{21} + M_{22}p_f)} \quad (5)$$

$$\text{With } (p_{0,f} = \sqrt{\frac{\varepsilon_0}{\mu_0}} n_{0,f} / \cos\theta_{0,f}).$$

Then, the transmittance can be expressed as [18]:

$$T = \frac{p_f}{p_0}|t|^2 \quad (6)$$

3. Results and discussion

In this section, the numerical results of the proposed periodic 1D PCs with superconductor photonic hetrostructure have been studied in the visible and near infrared (IR) wavelength range. The structure consists of the pair superconductor layers of K_3C_{60} [34] with (d_1 = 70nm, T_c = 19.5 K, λ_0 = 420 nm) and $HgBa_2Ca_2Cu_3O_{10}$ [35] with (d_2 = 10 nm, T_c = 135K, λ_0 = 177 nm) , respectively. Where, d_1 , d_2, Tc and λ_L are the thicknesses, the critical temperature and the London penetration depth of consider superconductor. Let us consider the proposed 1D superconducting PC that are delimited by dielectric cap layered with [n_3=2.5, d_3=20nm] and surrounded by air and glass substrate.

Figure 2 shows the dependence of the reflectance on the normalized frequency of the periodic superconducting PC consisting of the pair superconductor layers (K_3C_{60} / $HgBa_2Ca_2Cu_3O_{10}$). It is obvious that the proposed heterostructure exhibit a spectrum with two symmetric narrow reflectance peaks. The first one called Fano resonance and the second peak with two side bands form a broad up ward reflectance valey called the EIR that took place at 1.11 and 1.55 normalized frequencies. Furthermore, the amplitude of the Fano resonance of the pair superconductor is reach to 80%. However, the EIR reach to 50% of reflection. The given Fano resonance and EIR are due to the destructive interference of a discrete state with a continuum one. The proposed structures can serve as optical switching and light modulator.

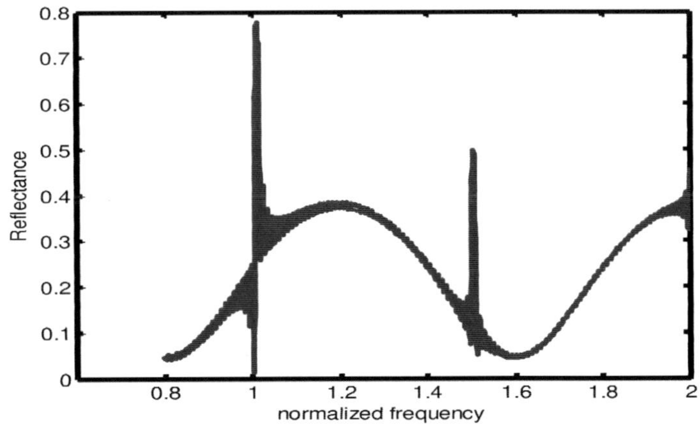

Figure 2: Dependence of the reflectance on the normalized frequency of the periodic superconducting PC consisting of the pair superconductor layers (K_3C_{60} / $HgBa_2Ca_2Cu_3O_{10}$).

Now, we investigated the influence of incident angle on the resonance properties of the periodic superconducting PC. Reflectance spectra versus incident angles for TE waves are shown in Figure 3. As the incident angle increases, both Fano resonance and EIR shift toward the longer frequencies for the proposed structure. This behavior is due to the two coupled oscillator modes that consider the EIR resonance as a special case of Fano resonance happened when the frequencies of strongly and weakly damped oscillators matched [33, 34].

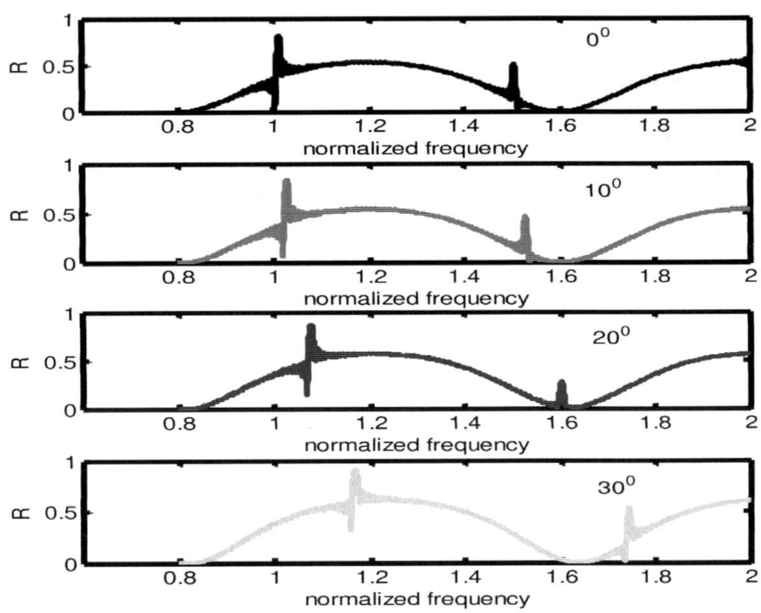

Figure 3: Reflectance spectra versus normalized frequency of the periodic superconducting PC for different incident angles.

Now, we discuss the influence of operating temperature on both Fano and EIR resonances. Figure 4 shows the dependence of the reflectance on the normalized frequency of the periodic superconducting PC for different value of operating temperature. Here the temperatures are set to be 2, 10, 15and 20K, respectively. It is obvious that the proposed structures both Fano and EIR peaks have a slowly increase of its amplitude of Fano resonance and the EIR resonance.

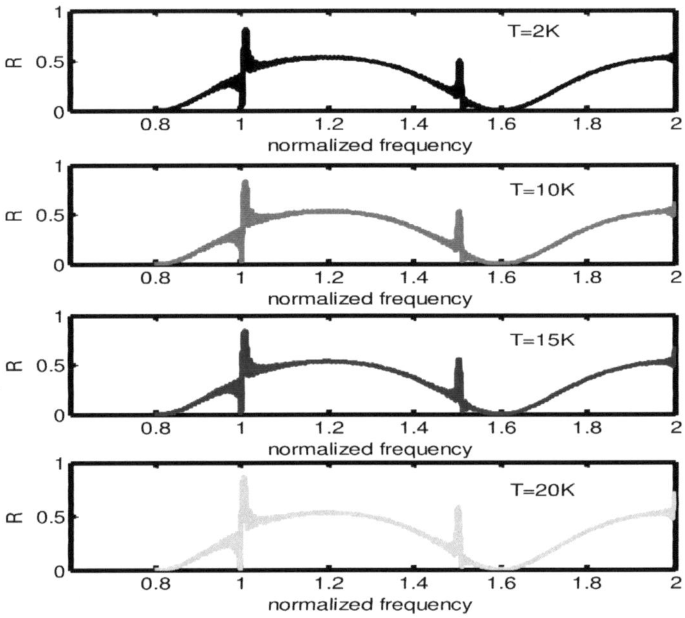

Figure 4: Reflectance spectra versus normalized frequency of the periodic superconducting PC at different values of the temperature.

We study the effect of change in thickness of the superconductor with thicknesses d_l of the periodic superconducting PC. In Figure 5, we have plotted the frequency-dependent reflectance for the superconducting PC in TE wave at different thicknesses d_l =70 nm, 90 nm, 100 nm and 150 nm respectively. By increasing the thickness d_l, the obtained Fano resonance of the periodic superconducting PCs shifted to lower normalized frequency region. Wherein, the EIR with symmetrical line shape shifted towards the high normalized frequency region accompanied by new Fano resonance peaks. It is obvious that from the spectrum the number of Fano resonance peaks depends on set thickness and can serve as multioptical switching. It constitute an advantages of macroscopic opto-electronic switching in which the damped of oscillation within this behavior is reduced.

The dependence of the reflectance for the TE polarized EMWs on the Fano resonance and EIR for the normal incidence angle is illustrated in Figure 6, for some refractive index of dielectric cap layer: 1.3, 1.6, 2 and 2.5.
By increasing the refractive index of dielectric cap layer, the reflectance exhibit a spectra in which position and amplitude of Fano resonance peak and the EIR decrease progressively for an increase of refractive index of its dielectric cap layer. This decrease in peak height means an incomplete resonance at the defect mode. For the refractive index of substrate layer, $\mathbf{n_s=2.5}$, the Fano resonance of the proposed superconducting PC become

symmetrical line shape. In this case, such dielectric cap layer is very important to calibrate the proposed structure by adjusting the characteristics of Fano resonance and EIR peaks.

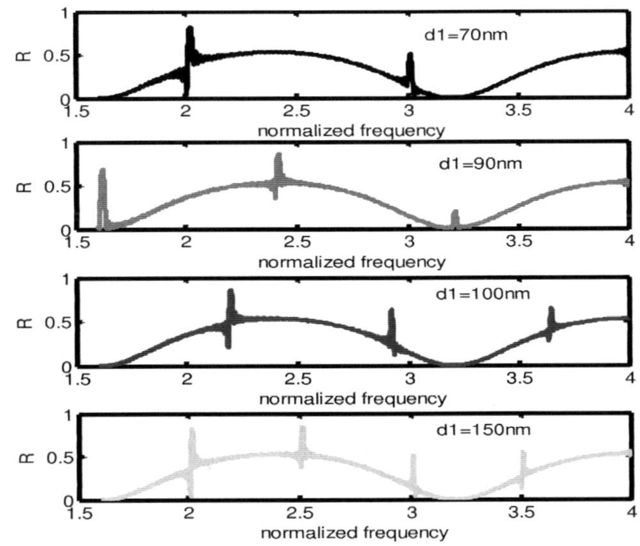

Figure 5: Reflectance spectra versus normalized frequency of the periodic superconducting PC at different values of thicknesses d_1 (nm).

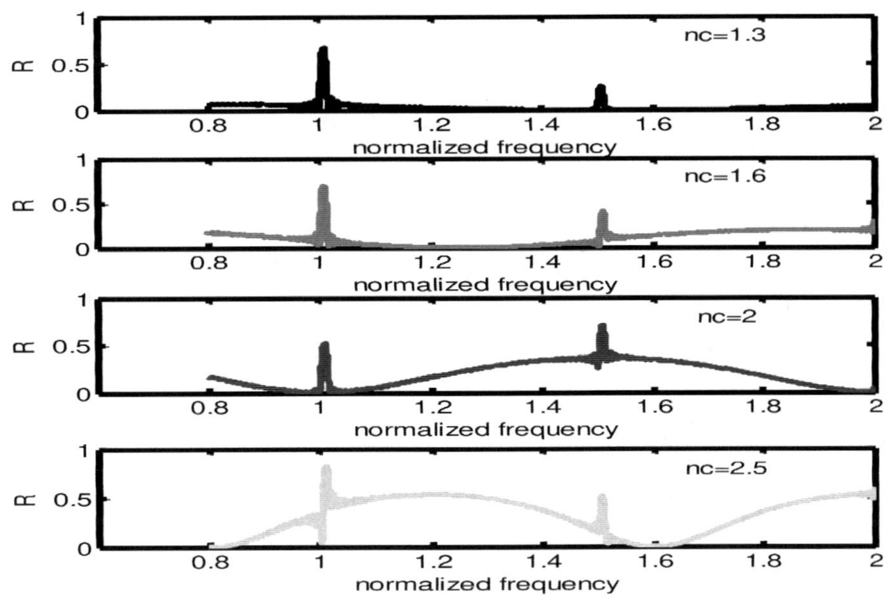

Figure 6: Reflectance spectra versus normalized frequency of the periodic superconducting PC at different values of refractive index of consider dielectric cap layer.

In Figure 7, the resonance properties are depicted as functions of normalized frequency for different thicknesses of dielectric cap layer dc (nm). By increasing the dc, the reflectance exhibits a spectrum in which the Fano resonance in the periodic superconducting PCs present two asymmetric resonance peaks. However, the position of given EIR peaks shifted to the lower normalized frequency region when we increase the thickness of dielectric cap layer. It should be noted that its peak of EIR become with asymmetric shape line from dc=8nm due to the increase of the phase shift in the two coupled oscillator mode.

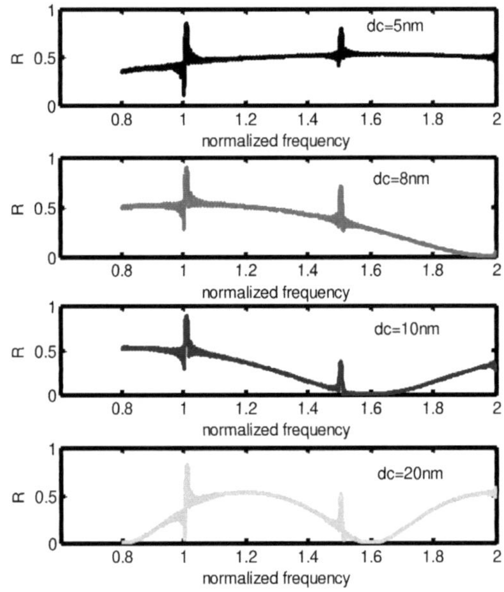

Figure 7: Reflectance spectra versus normalized frequency of the periodic superconducting PC at different values of thicknesses of dielectric cap layer dc (nm).

Conclusion:

A theoretical study of Fano resonance and EIR properties in periodic 1D superconducting PCs consisting of (K_3C_{60} / $HgBa_2Ca_2Cu_3O_{10}$) superconducting materials are investigated. The suggested structure exhibits a tunable Fano resonances and EIR resonance peak accompanied by asymmetrical line shape very sensitive to thickness and refractive index of dielectric cap layer, constituent materials, dependence incidence angle and temperature. Furthermore, the obtained spectra are tunable by adjusting the value of the optical parameters. As a result, a good response with maximum peaks of reflection is given by the proposed structure. Finally, the structure will be helpful in the field of optical communication and it can be used as optical switch.

4. References

[1] Yablonovitch, E. (1987). Inhibited spontaneous emission in solid-state physics and electronics. Phy. Rev. Lett., 58, 2059.

[2] Yablonovitch,E. (1993). Photonic band-gap structures. JOSA B, 10, 283-295.

[3] Chen, X.D., Shi, F.L., Iu, H., Lu, J.-C., Deng, W.-M., Dia, J.Y., Cheng,Q., & Dong, J.-W. (2018). Tunable Electromagnetic Flow Control in Valley Photonic Crystal Waveguides. Phys. Rev. Applied, 10, 044002.

[4] Aly, A. H., & Sayed, F. A, (2020). THz cutoff frequency and multifunction Ti2Ba2Ca2Cu3O10/GaAs photonic bandgap materials. Int. J. Mod. Phy. B, 34(10), 2050091.

[5] Aly, A. H., Mohamed, D., Mohaseb, M.A., Abd El-Gawad, N. S., & Trabelsi, Y., (2020). Biophotonic sensor for the detection of creatinine concentration in blood serum based on 1D photonic crystal. RSC Advances, 10, 31765 – 31772.

[6] Kabir, M.A., Hassan, M.M., Ahmed, K., M.S.Mani Rajan, Aly, A. H., Hossain, N., & Bikash Paul, B., K., (2020). Novel spider web photonic crystal fiber for robust mode transmission applications with supporting orbital angular momentum transmission property. Opt. Quant. Electron., 52:331.

[7] Aly, A. H., Mohamed, D., & Mohaseb, M.A., (2020). Metamaterial Control of Hybrid Multifunctional High-Tc Superconducting Photonic Crystals for 1D Quasi-periodic Structure Potential Applications. Materials Research, 23(3): e20190695.

[8] Zaky, Z.A., & Aly, A. H., (2020). Theoretical Study of a Tunable Low-Temperature Photonic Crystal Sensor Using Dielectric-Superconductor Nanocomposite Layers. J Supercond Nov Magn., 33:2983–2990.

[9] Aly, A. H., Mohamed, D., & Mohaseb, M.A., (2019). Theoretical and simulation study in defective semiconductor layer that incorporated with superconducting-dielectric photonic crystal. Int. J. Mod. Phy. B, 33(32), 1950397.

[10] Aly, A. H., Ameen, A.A., & Vigneswaran, D. (2019). Superconductor Nanometallic Photonic Crystals as a Novel Smart Window for Low-Temperature Applications. J Supercond Nov Magn., 32(2), 191–197.

[11] Trabelsi, Y., Ali, N.B., Aly, A. H., & M. kanzari M. (2020). Tunable high Tc superconducting photonic band gap resonators based on hybrid quasi-periodic multilayered stacks. Physica C, 576, 1353706.

[12] Wiesmann, S., Bergenek, K., Linder, N., & Schwarz, U.T., (2009). Photonic crystal LEDs – designing light extraction. Laser and Photon. Rev. 3, 262–286.

[13] Nozaki, K., & Baba,T., (2004). Quasiperiodic photonic crystal microcavity lasers. Appl. Phys. Lett. , 84, 4875-4877.

[14] Gao, Y., Huang, X., & Xu, X., (2014) Electro-optic modulator based on a photonic crystal slab with electro-optic polymer. Optics Express, 22, 8765-8778.

[15] Zaky, A.Z., Ahmed, A.M., Shalaby, A.S., & Aly, A.,H., (2020). Refractive index gas sensor based on the Tamm state in a one-dimensional photonic crystal: Theoretical optimization. Sci. Rep., 10, 9736.

[16] Nouman, W.M., Abd El-Ghany, A.E.-S., Sallam, M.,S., Dawood, A.B., & Aly, A. H. (2020). Biophotonic sensor for rapid detection of brain lesions using 1D photonic crystal. Opt Quant Electron., 52, 287.

[17] Trabelsi, Y., N. Ben Ali, N.B., M. & Kanzari, N., (2019). Tunable Narrowband optical filters using superconductor / dielectric generalized Thue-Morse photonic crystals. Microelectronic Engeneering, 213, 41–46.

[18] Aly, A. H., Mohamed, D., Elsayed H., & Mehaney, A. (2018). Fano-resonance by means of the one-dimensional superconductor photonic crystals. J. Supercond. Nov. Magn., 31, 3827–3833.

[19] Aly, A. H., & Mohamed, D. (2019). The Optical Properties of Metamaterial-Superconductor Photonic Band Gap With/Without Defect Layer. J Supercond. Nov Magn, 32(7), 1897–1902.

[20] Aly, A. H., Abdel Ghany, S.E.-S., Kamal, B. M., & Vigneswaran, D. (2020). Theoretical studies of hybrid multifunctional YaBa2Cu3O7 photonic crystals within visible and infra-red regions. Ceramics International, 46.1, 365-369.

[21] Trabelsi , Y., (2019). Output multichannel optical filter based on hybrid photonic quasicrystals containing a high-Tc superconductor. Photonics and Nanostructures - Fundamentals and Applications ,36 ,100724.

[22] Aly, A. H., Ryu, S.W., Hsu, H.T., & Wu , A.J. (2009). THz transmittance in one-dimensional superconducting nanomaterial-dielectric superlattice. Materials Chemistry and Physics, 113, 382-384.

[23] Trabelsi , Y., Belhadj,W., Benali, N., & Kanzari, M., (2019). Photonic Band Gap Properties of One-dimensional Generalized Fibonacci Photonic Quasicrystal Containing Superconductor Material. J. Supercond. Nov. Magn. 32 (11), 3541-3547.

[24] Aly, A.H., & Sayed, W., (2016). Superconductor-Semiconductor metamaterial photonic crystals. J. Supercond Nov Magn, 29:1981–1986.

[25] Elsayed, H.A., & Aly, A.H., (2018). Terahertz Frequency superconductor-nanocomposite photonic band gap. Intrenational Journal of Modern Physics B, 32(5) 1850056.

[26] Aly, A. H., & Mohamed, D., (2015). BSCCO/SrTiO3 One Dimensional Superconducting Photonic Crystal for Many Applications. J Supercond . Nov Magn, 18, 1699-1703.

[27] Segovia-Chaves,F., Vinck-Posada, H., Trabelsi , Y., & Ben Ali , N. (2020). Transmittance spectrum in a one-dimensional photonic crystal with Fibonacci sequence superconductor–semiconductor. Optik ,217,164803.

[28] Aly, A. H., (2008). Metallic and superconducting photonic crystal. J Supercond . Nov Magn., 21:421-425.

[29] Athe, P., & Srivastava ,S., (2015). Tunable Fano Resonance in One-Dimensional Superconducting Photonic Crysta. J Supercond . Nov Magn., 28,2331–2336.

[30] Song, J., Zaccaria, R.P., Yu, M., B., & Sun, X., W. (2006). Tunable Fano resonance in photonic crystal slabs. Optics Express, 14, 8812-8826.

[31] Singh, R., Al-Naib, I.A., Koch, M.,& Zhang, W. (2011). Sharp Fano resonances in THz metamaterials. Opt. Express, 19, 6312–6319.

[32] Dong, G., Wang Y., & Zhang, X., (2018). High-contrast and low-power all-optical switch using Fano resonance based on a silicon nanobeam cavity. Optics Letters, 43, 5977-5980.

[33] Luk'yanchuk, B., Zheludev, N.I., Maier, S.A., Halas, N.J., Nordlander, P., Giessen, H., & Chong, C., T., (2010). The Fano resonance in plasmonic nanostructures and metamaterials. Nat. Mater., 9(8), 707–715.

[34] Srivastava, S.K., (2014). Study of Defect Modes in 1d Photonic Crystal Structure Containing High and Low Tc Superconductor as a Defect Layer. J Supercond . Nov Magn, 27, 101–114.

[35] Rahimi , H., (2016). Analysis of photonic spectra in Thue-Morse, double-period and Rudin-Shapiro quasiregular structures made of high temperature superconductors in visible range. Optical Materials, 57,264-271.

IOP Conf. Series: Materials Science and Engineering **956** (2020) 012005 doi:10.1088/1757-899X/956/1/012005

Simulation of CuO/ZnO heterojunction for photovoltaic applications

Mahmoud Abdelfatah [1], Nagi M. El-Shafai [2], Walid Ismail [1], Ibrahim M. El-Mehasseb [2] , and Abdelhamid El-Shaer [1]

[1] Physics Department, Faculty of Science, KafrelSheikh University, KafrelSheikh, 33516, Egypt.

[2] Nanotechnology center, Chemistry Department, Faculty of Science, Kafrelsheikh University 33516, Egypt

E-mail: mahmoud.abdelfatah@sci.kfs.edu.eg

Abstract. Inorganic semiconductors become one of the most gifted candidate for photovoltaic applications instead of high cost silicon devices and organics ones. Therefore, in this paper, inorganic CuO/ZnO heterojunction is simulated as thin layer solar cells employing SCAPS program. Both effect of CuO and ZnO layers thickness on basic parameters of solar cells is investigated deeply. Simulated results of this solar cell device indicate that the absorber layer is responsible for separation process of generated carriers. Higher efficiency of about 23 % could be achieved with 2 μm of CuO thickness and 100 nm of ZnO layer for devices. Generally, the main aspects that controlling the solar cell performance are depletion region width, built-in potential, the charge carrier collection length and generation and recombination rates

1. Introduction

Although Popular photovoltaics such crystalline silicon (c-Si), cadmium telluride (CdTe), and copper indium gallium diselenide (CIGS) have higher efficiencies, but there are some disadvantages of them [1]. For example, the fabrication of c-Si needs high temperature and ultra-high vacuum, which make such process, is very expensive. CIGS and CdTe photovoltaics are based on non-earth abundant materials [1, 2].

Therefore, the wide spread of solar cells requires some aspects such as the used materials for fabrication should be abundant, inexpensive and non-toxic. In addition, the solar cells should easily fabricated at low temperature and cost-effective techniques [1-4].

For such purpose, inorganic metal oxide semiconductors are the most promising candidates which nearly imply these requests and considered as an alternative to the high cost silicon devices and organics ones [1, 2, 5-7].

Zinc oxide (ZnO) is a promising applicant for solar cell applications [1-3, 5, 8-12]. ZnO is a wide band gap semiconductor (3.37 eV) with negative conductivity [11, 13]. Moreover, ZnO has various application in many fields such light-emitting diodes [5, 6] catalysts [14], antioxidants [15], photodetectors [16] and water treatment [15].There are different techniques could be employed to have several shapes of ZnO nanostructures such as chemical vapor deposition [17],pulsed laser deposition [18], molecular beam epitaxy [19], sputtering [1, 2], chemical bath deposition [20], electrodeposition [21], and hydrothermal [5, 6, 11].

Combination of ZnO nanostructures with another p- type semiconductor is a key to have higher efficiency heterojunction solar cells. Copper oxide (CuO) considered one of these semiconductors to fabricate solar cells since it has property band gap (1.2 eV) for such applications [22]. CuO has higher light coefficient with lower thermal emittance. Moreover, less mismatch between lattices will be achieved through ZnO/CuO heterojunction [22]. Consequently, alignment of energy level will develop the separation process of charge carriers and therefore increase the solar cells performances [23].

Numerical study of nanomaterials have become powerful methods for scientists and engineers to predicate the efficiency and performance of such materials on optoelectronic applications like solar cells, and light emitting diodes. Here, we deeply investigated numerical simulation of ZnO/CuO heterojunction as solar cells. The effect of active layer and window layer thickness on the solar cell basic parameters was calculated employing Solar Cell Capacitance Simulator Structures (SCAPS) program in order to know the optimal thickness to obtain higher performances [24, 25].

2. Materials and Methods

2. 1 Solar Cell Structure:

The structure of simulated solar is presented in figure 1.In this case, CuO film was employed as an active layer and ZnO was used as window layer in the presented heterojunction solar cell. Gold (Au) and Florin doped tin oxide (FTO) were employed as back and front contacts, respectively.

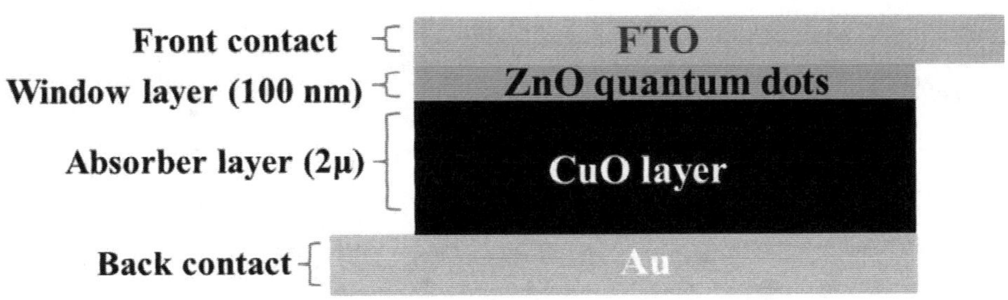

Figure 1. Diagram structure of CuO/ZnO heterojunction thin film solar cell.

2.2 Materials Properties and Simulation process:

The properties of layers, which are used in numerical simulation, are obtained from previous literature[22, 26] are listed table 1.

IOP Conf. Series: Materials Science and Engineering **956** (2020) 012005 doi:10.1088/1757-899X/956/1/012005

Table 1. Properties of Materials used in our simulation

Material properties	p-CuO	n-ZnO
Relative permittivity	18.1	86
Bandgap (eV)	1.2	3.37
Electron affinity (eV)	4.07	3.9
Effective density of states of valence band maximum(cm^{-3})	5.5E+20	1.50E+21
Effective density of states of conduction band minimum (cm^{-3})	3E+19	1.32E+20
Acceptor concentration(cm^{-3})	1.00E+16	0
Donor concentration(cm^{-3})	0	1.00E+19
Mobility of Hole (cm^2/V s)	10	20
Mobility of Electron (cm^2/V s)	0.1	10

In this paper, Solar Cell Capacitance Simulator structures (SCAPS-1D) software program was used to simulate and explore devices [27-29] as shown in figure2 as screenshot from the program. The impact of thickness of window layer as well as active layer on solar cell basic parameters was examined deeply. One sun illumination (AM 1.5 spectrum illumination -100 mW/cm^2) was used for all numerical simulations calculations while keeping temperature constant at 300 K. Both Poisson and continuity equations of holes and electrons were employed for the simulations [23]:

$$\frac{d^2}{dx^2}\Psi(x) = \frac{e}{\varepsilon_0\varepsilon_r}\left(p(x) - \mathrm{n(x)} + N_D - N_A + \rho_p - \rho_n\right)(1)$$

$$\frac{dJ_n}{dx} = \mathrm{G} - \mathrm{R} \qquad \text{and} \qquad \frac{dJ_p}{dx} = \mathrm{G} - \mathrm{R} \qquad (2)$$

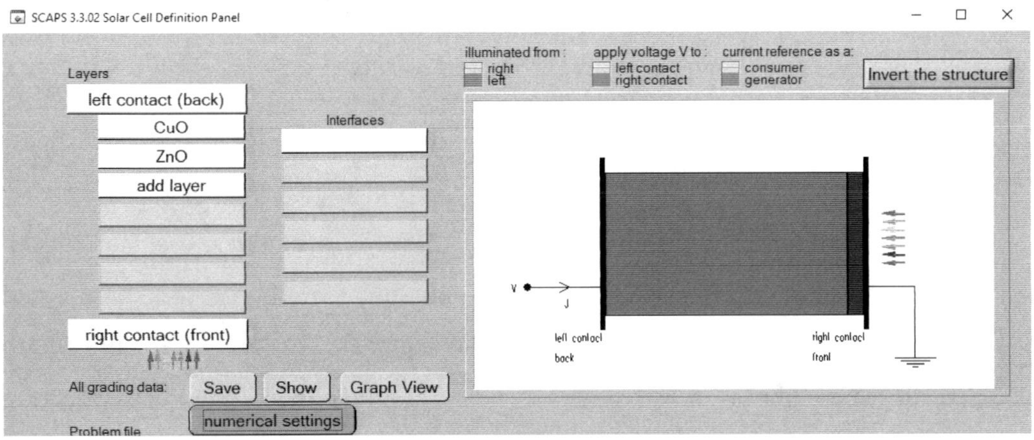

Figure 2. Screenshot from SCAPS program for simulated CuO/ZnO solar cell.

3. Results and discussion

To investigate the influence of absorber layer thickness on solar cell basic parameters (open circuit voltage (Voc),short circuit current density (Jsc), fill factor (FF),as well ascell efficiency (η)). The CuO layer thickness was varied from 0.01 to 5 μm while the window layer was fixed at 100 nm.

Figure 3 shows the outline of the J_{sc}, V_{oc}, FF, and η versus p-CuO film thickness. It is obvious that that V_{oc} rises sharply from 0.715 to 0.785 V with increasing of the absorber layer as illustrated infigure3a.This behavior could be attributed to the enhancement of built-in potential which produced by diffusion of holes from active layer to window layer of the solar cell [1]. This outcome is the same as that reported for Cu_2O solar cells in which increasing thickness creates a built-in potential with maximum value of about 0.7 V[1, 30]. Also, the increasing of V_{oc} could be explained by increasing the charge carriers generation rate as well as

IOP Conf. Series: Materials Science and Engineering **956** (2020) 012005 doi:10.1088/1757-899X/956/1/012005

reducing leakage current in interface of heterojunction which supported by the equation[1].

$$V_{oc} = \frac{k_B \mathrm{T}}{\mathrm{q}} \ln\left(1 + \frac{I_{ph}}{I_0}\right) \quad (3)$$

Figure 3b clarifies J_{sc} of simulated heterojunction solar cells. It is clear that the current increase from about 36.4 mA/cm^{2}to 38.7 mA/cm^2 with thickness growing from 100 nm to 500 nm. With further increase in layer thickness, the current become constant at about 38.7 mA/cm^2. The length of charge collection could recognize improvement of Jsc with thickness increasing of CuO layer from 100 to 500 nm where photogenerated carriers will recombine together before collecting with lower thickness while with higher thickness; they will be separated and gathered easily. So generation rate of hole electron pairs will be improved with thickness increasing up to that determined from the absorption coefficient of CuO layer [30] achieving constant value of J_{sc} .

Figure 3c displays the FF behavior where a partial increase with thickness from 100 nm to 500 nm followed by monotonically decrease with thickness increase. Such change could be deduced from FF equation, which shows that FF is inverse proportional with V_{oc} and J_{sc} values as following.

$$\mathrm{FF} = \frac{V_m J_m}{V_{oc} J_{sc}} \quad (4)$$

IOP Conf. Series: Materials Science and Engineering **956** (2020) 012005 doi:10.1088/1757-899X/956/1/012005

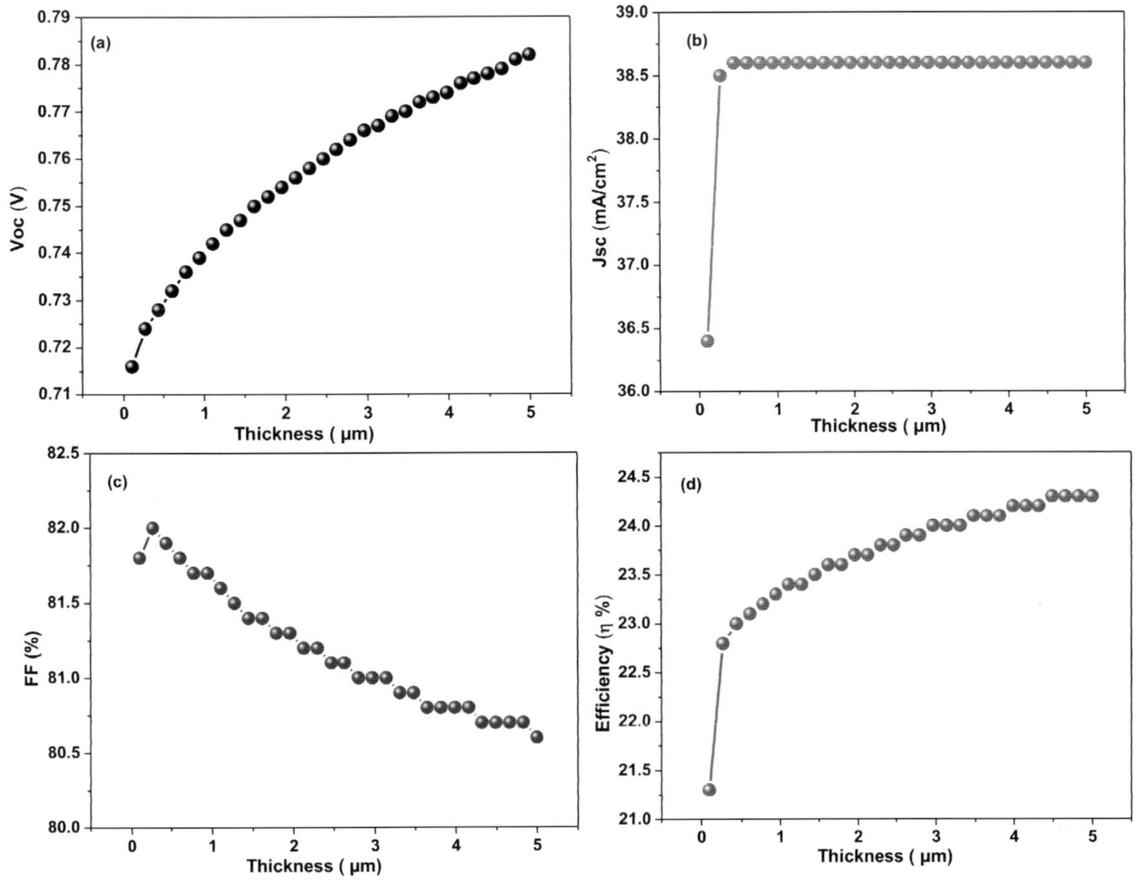

Figure 3. Simulated (a) V_{oc}, (b) J_{sc}, (c) FF, and (d) η of ZnO/CuO heterojunction devices with absorber layer thickness.

Figure 3d shows the change in η of simulated devices with thickness of CuO layer. It is obvious that η displays significant improvement with thickness increasing from 100 nm to 500 nm followed by slight increase from about 21.25 % to about 23 % with increasing thickness up to 5 μm.

Such trend of the η could be explained by its dependence on the V_{oc},, J_{sc}, and the FF values as presented by the following equation [31]:

$$\eta = \frac{FF\, V_{oc}\, J_{sc}}{P_{in}} \qquad (5)$$

From that, it could be assume that 2-μm thickness for CuO film is the optimum one to have higher performance of the simulated photovoltaic solar cell here

Figure 4 display a brief for the simulated results of basic parameters for the heterojunctions with different window layer thicknesses keeping the absorber layer constant at 2 μm. It is clear from figure 4a that, V_{oc} decline with ZnO layer increasing. This case happen because of reducing of photogenerated current and increasing of the recombination rate of charge carries according to equation (3). In addition, increasing thickness of window layer blocks light to pass through the heterojunction interface. Moreover, this confirmed that the built-in potential which is responsible for the formation of V_{oc}, is formed in active layer side (CuO) [1]. Figure 4b presents the J_{sc} where it has the same behavior the V_{oc}. This style could be explained by increasing the leakage current, which affect negatively on the generated current as well as length of charge collection reached maximum value. Inserting of buffer layers at the interface of the heterojunction could paly important factor to reduce the leakage current through reducing layer mismatch and increasing band gaps alignment.

The behavior of FF as illustrated in figure 4c have inverse shapes of the J_{sc} and the V_{oc} where it is inverse proportional to their values as previously discussed according to equation (4). Figure 4d reflects the dependence of the η on values of the V_{oc}, the J_{sc}, and the FF as shown in equation (5). In this situation, then firstly increase from 22.9 to 23.3 % by increasing thickness from 10 nm to 100 nm. Then, with further thickness increasing, it falls down to about 23.1% and become constant. It can conclude that the higher performance of solar cells could be obtained with 100 nm thickness of ZnO as window layer.

IOP Conf. Series: Materials Science and Engineering **956** (2020) 012005 doi:10.1088/1757-899X/956/1/012005

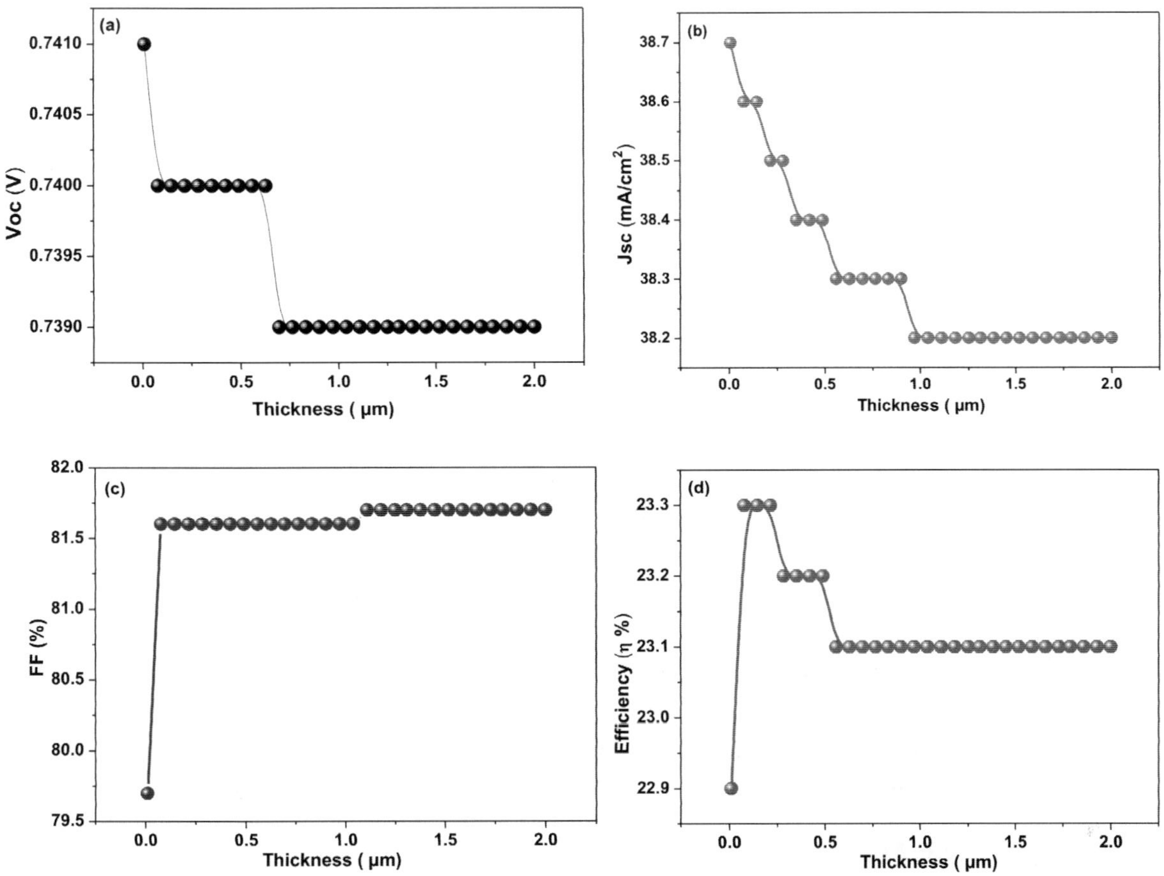

Figure 4. The calculated (a) V_{oc}, (b) J_{sc}, (c) FF, and (d) η of the heterojunction devices versus the window film thickness.

4. Conclusion

CuO/ZnO heterojunction solar cells were simulated employing SCAPS program. Impact of active layer and window layers on solar cell basic parameters was examined deeply. Results of simulated devices showed that the built-in potential was formed in CuO side and higher efficiency of 23 % was achieved with thickness of 2 μm of absorber layer. The thickness of 100 nm for ZnO as window layer should applied to have higher efficiencies. The main factors that directing the performance of heterojunction devices are the charge carrier collection length, generation and recombination rates, depletion region width and built-in potential. Application of our simulated results on the experimental area would support and improve the solar cell conversion efficiency.

Acknowledgments

Authors are thankfully for Prof. Dr. Marc Burgelman for SCAPS program providing. This paper is based upon work supported by Science, Technology & Innovation Funding Authority (STDF) under grant (33389) as well as by the Academy of Scientific Research and Technology (ASRT) (Project ID: 6689).

References

[1] Abdelfatah M, Ledig J, El-Shaer A, Wagner A, Marin-Borras V, et al. 2016. *Solar Energy Materials and Solar Cells* 145: 454-61.
[2] Abdelfatah M, Ledig J, El-Shaer A, Wagner A, Sharafeev A, et al. 2015. *Solar Energy* 122: 1193-8.
[3] Abdelfatah M, Ledig J, El-Shaer A, Sharafeev A, Lemmens P, et al. 2016. *ECS Journal of Solid State Science and Technology* 5: Q183-Q7.
[4] Lee YS, Heo J, Siah SC, Mailoa JP, Brandt RE, et al. 2013. *Energy & Environmental Science* 6: 2112-8.
[5] Abdelfatah M, Ismail W, El-Shaer A. 2018. *Materials Science in Semiconductor Processing* 81: 44-7.
[6] El-Shaer A, Ismail W, Abdelfatah M. 2019. *Materials Research Bulletin* 116: 111-6.
[7] El-Shafai NM, Abdelfatah MM, El-Khouly ME, El-Mehasseb IM, El-Shaer A, et al. 2020. *Applied Surface Science* 506: 144896.
[8] Baek SK, Lee KR, Cho HK. 2013. *Journal of Nanomaterials* 2013: 6.
[9] El-Shaer A, Tadros M, Khalifa M. 2015. *Nat. Sci.* 13: 14-22.
[10] Jeong S, Mittiga A, Salza E, Masci A, Passerini S. 2008. *Electrochimica Acta* 53: 2226-31.
[11] Abdelfatah M, El-Shaer A. 2018. *Materials Letters* 210: 366-9.
[12] El-Shaer A, Abdelfatah M, Mahmoud KR, Momay S, Eraky M. 2020. *Journal of Alloys and Compounds* 817: 152799.
[13] Young S-J, Chiou C-L, Liu Y-H, Ji L-W. 2016. *Inventions* 1: 3.
[14] Cao M, Wang F, Zhu J, Zhang X, Qin Y, Wang L. 2017. *Materials Letters* 192: 1-4.
[15] El-Shafai NM, Shukry M, El-Mehasseb IM, Abdelfatah M, Ramadan MS, et al. 2020. *Materials Science and Engineering: B* 259: 114596.

[16] Bedia A, Bedia FZ, Benyoucef B, Hamzaoui S. 2014. *Physics Procedia* 55: 53-60.
[17] Zhong J, Muthukumar S, Saraf G, Chen H, Chen Y, Lu Y. 2004. *Journal of Electronic Materials* 33: 654-7.
[18] Hong J-I, Bae J, Wang ZL, Snyder RL. 2009. *Nanotechnology* 20: 085609.
[19] El-Shaer A, Bakin A, Al-Suleiman M, Ivanov S, Mofor AC, Waag A. 2007. *Superlattices and Microstructures* 42: 129-33.
[20] El-Shaer A, Abdelfatah M, Basuni A, Mosaad M. 2018. *Chinese journal of physics* 56: 1001-9.
[21] El-Shaer A, Abdelwahed AR, Mosaad M. 2015. International Journal of Research in Engineering and Science, 3 : 28-36.
[22] Zhu L, Shao G, Luo J. 2011. *Semiconductor science and technology* 26: 085026
[23] Abdelfatah M, Ismail W, El-Shafai NM, El-Shaer A. 2020. *Materials Technology*: 1-9.
[24] Nguyen DL. 2020. *Engineering Research Express* , **2**: 025033.
[25] Parashar D, Krishna VG, Moger SN, Keshav R, Mahesha M. 2020. Transactions on Electrical and Electronic Materials, **13**: 0123456789.
[26] Zhu L, Shao G, Luo J. 2013. *Semiconductor Science and Technology* 28: 055004.
[27] Ouédraogo S, Zougmoré F, Ndjaka J. 2013. *International Journal of Photoenergy* 2013.
[28] Djinkwi Wanda M, Ouédraogo S, Tchoffo F, Zougmoré F, Ndjaka J. 2016. *International Journal of Photoenergy* 2016.
[29] Burgelman M, Nollet P, Degrave S. 2000. *Thin Solid Films* 361: 527-32.
[30] Musselman KP, Marin A, Schmidt-Mende L, MacManus-Driscoll JL. 2012. *Advanced Functional Materials* 22: 2202-8.
[31] Baloch AA, Aly SP, Hossain MI, El-Mellouhi F, Tabet N, Alharbi FH. 2017. *Scientific reports* 7: 11984.

IOP Conf. Series: Materials Science and Engineering **956** (2020) 012006 doi:10.1088/1757-899X/956/1/012006

Phase transition of $Cd(OH)_2$ and physical properties of CdO microstructures prepared by precipitation method for optoelectronic applications

Walid Ismail*, Abdelhamid El-Shaer, Mahmoud Abdelfatah

*Physics Department, Faculty of Science, KafrelSheikh University, KafrelSheikh, 33516, Egypt.

E-mail address: walid_s2@yahoo.com

Abstract

Low cost precipitation technique is employed here to have CdO microstructures as a promising material for optoelectronic applications. Cadmium nitrate hexahydrate and sodium hydroxide were mixed together to have white precipitate of cadmium hydroxide. The product white precipitate then dried at 100 ^{o}C followed by different annealing temperatures (200, 300, 400 and 700 ^{o}C) for 2 h. The formed samples are characterized employing XRD, SEM, and TEM techniques to investigate the structural and morphology properties. The XRD results indicate that as prepared exhibited hexagonal β-$Cd(OH)_2$ phase. For annealed samples at 200, mixed phases of γ-$Cd(OH)_2$ hexagonal as well as monoclinic were appeared. γ-$Cd(OH)_2$ monoclinic phase and CdO cubic phase were formed with calcined temperature of 300°C. Pure cubic phase of CdO crystal structure was obtained at 400°C. With higher annealed temperature of 700°C, both cubic phases of CdO and CdO_2 are appeared. The results designate that CdO microstructures formed with single-phase formation of cubic structure at 400°C, which is the most advantageous for optoelectronic applications.

1. Introduction

Recently, semiconductor metal oxides have expected highly attention because of its unique properties, which give them high prospects in many fields and various applications [[1]-[8]]. Among these oxides, cadmium oxide (CdO) is a semiconductor with negative conductivity, unique electrical, optical, and catalytic properties [[9]]. CdO has direct band gap that could be varied from 2.2 to2.7 eV [[10]].

Lately, numerous researchers have concentrated on CdO because of various applications in optoelectronic especially solar cells [[11],[12]], photodiodes [[13]], phototransistors [[15]], transparent electrodes and gas sensors [[16]]. CdO could be synthesizes in diverse morphologies such as rhombus [[14]], nanoparticles [[17]], nanocrystals [[19]], nanosticks [[20]], nanowires [**Error! Reference source not found.**], nanoclusters [[22]], and nanocubes [22]]. Such shapes could achieved empolying various techinqies namely sonochemical [[24]], solvothermal [[24]], microwave [[24]], evaporation [[9]], chemical method [**Error! Reference source not found.**], and thermal techniques[[24]]. $Cd(OH)_2$ as well as CdO nanostructure have been grown by solvothermal where $Cd(OH)_2$ nanoparticles transformation to CdO microstructures was taken place at 500 ^{o}C [[27]].

In the present work, solvothermal rout was employed to synthesis $Cd(OH)_2$. The impact of calcination temperature on formed sample was investigated deeply. The optical and the structural were characterized employing XRD, TEM, and SEM whereas stability and wet loss was measured by TGA.

2. Experimental

Sodium hydroxide [NaOH] and Cadmium nitrate hexahydrate [$Cd(NO_3)_2$ -$6H_2O$], with 99% purity were bought from Merck. Typically, of $Cd(NO_3)_2$ -$6H_2O$ with molarity of 0.5 M was prepared then dropwise of NaOH (2 M) was added under stirring. Stirring process was continually until the mixture become a white precipitate of Cadmium hydroxide hexahydrate. The formed precipitate then filtered and washed with deionized water for 3 times to remove any excess of the precursors as well as impurities from product. Further, the precipitate was fine dried for 4 h at 100°C where nanoparticles of cadmium hydroxide was achieved. Finally, different annealing temperatures (200, 300, 400 and 700 ^{o}C) were performed for 2 h in order to probe impact of calcination temperature on formed samples [[28]].

3. Results and Discussion

The crystal structure of fabricated as well as calcined samples was examined by XRD as displayed in Figure (1). The sample as prepared exhibited hexagonal **β**-$Cd(OH)_2$ nanoparticles and cube-like $Cd(OH)_2$ where diffraction planes of such phases were appeared clearly as pervious reported [28]]. XRD pattern is the same of the card No. (JCPDS 31-228) that is confirmed formation of $Cd(OH)_2$ hexagonal phase [[24]]. It is clear that the samples annealed at 200 ^{o}C and 300 ^{o}C exhibited diffraction peaks corresponding to the (130), (031), and (310) planes which matched to γ-$Cd(OH)_2$ monoclinic phase (JCPDS: 20-0179). Moreover, planes of CdO cubic phase was appeared. For further high temperatures annealing at of 400 ^{o}C, pure CdO cubic crystal structure was formed where all the diffraction peaks of such phase were performed as in card No. (JCPDS: 65-2908). The observed peaks at 2θ values at around 32.994^{o}, 38.279^{o}, 55.265^{o}, 65.820^{o} and 69.239^{o} were corresponded to (111), (200), (220), (311) and (222) planes, respectively. Very

IOP Conf. Series: Materials Science and Engineering **956** (2020) 012006 doi:10.1088/1757-899X/956/1/012006

strong (111) orientation plane was observed for as manufactured nanoparticles where this result with agreement with results reported previously [[29]] [[30]]. For the sample that calcined at 700 °C, another peaks at 2θ values of about 44°, 64.3°, and 77.5° were appered which belongs to (220), (023) and (331) planes for CdO_2 cubic phase.

This could be explained as increasing annealing temperature will lead to the probability of oxidation process that cause the formation of cubic phase of CdO_2 [[30]]. Therefore, to have pure single phase of CdO, the $Cd(OH)_2$ must annealed at 400 °C.

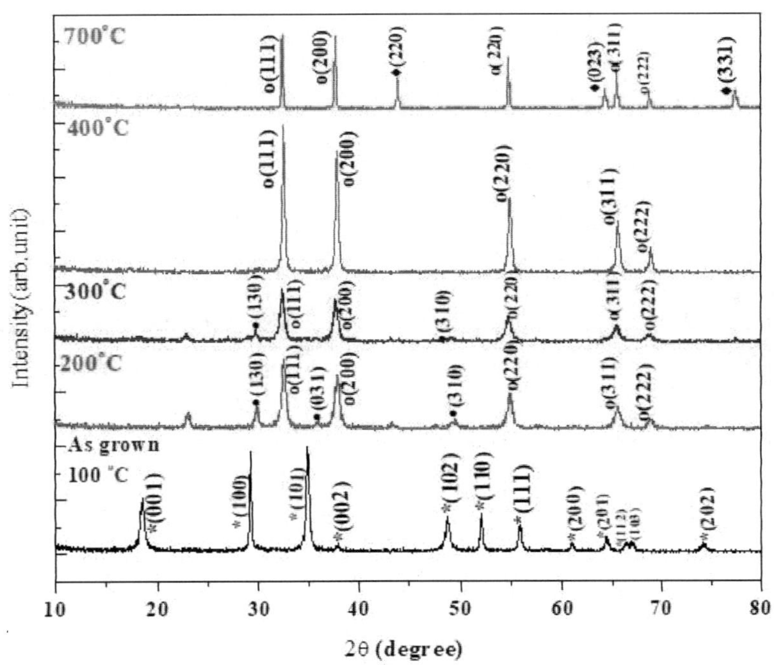

Figure 1. XRD of as grown and calcined $Cd(OH)_2$ and CdO nanoparticles ($Cd(OH)_2$; *Hexagonal- • Monoclinic); (CdO; ° Cubic)- CdO_2;♦ Cubic).

Scanning electron microscopy (SEM) results of samples that calcined at 400 and 700 °C of CdO microstructures is presented in the Fig. 2. The results show microstructures have been constructed in sphere shape where particles are congregated to produce polydispersive clusters. The clusters in sample annealed at 400 °C is very condensed with very small particle size, whereas; particle size of sample annealed at 700 °C is larger and the spherical shape becomes clear. That enlarge of cluster size due to heat treatments and increase of agglomeration [[31]].

Figure 2.SEM images of CdO nanocrystals samples after thermal treatment at 400°C and 700 °C.

The TEM images of CdO microstructures of samples that calcined at 400 and 700 °C of CdO nanocrystals are shown in the Fig. 3. The TEM image displays morphology of spherical like shapes with average size of particle of about 151 nm with annealed sample at 400 °C. On the other side, the sample calcined at 700 °C has average particle size of about 173 nm. The particles are spherical and elliptical or rice-shaped microstructures [**Error! Reference source not found.**].

IOP Conf. Series: Materials Science and Engineering **956** (2020) 012006 doi:10.1088/1757-899X/956/1/012006

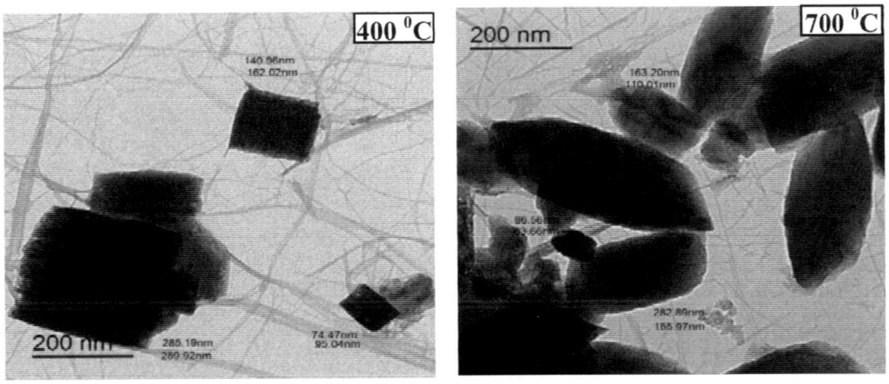

Figure 3. TEM images of CdO microstructures annealed at 400 ^{o}C and 700 ^{o}C.

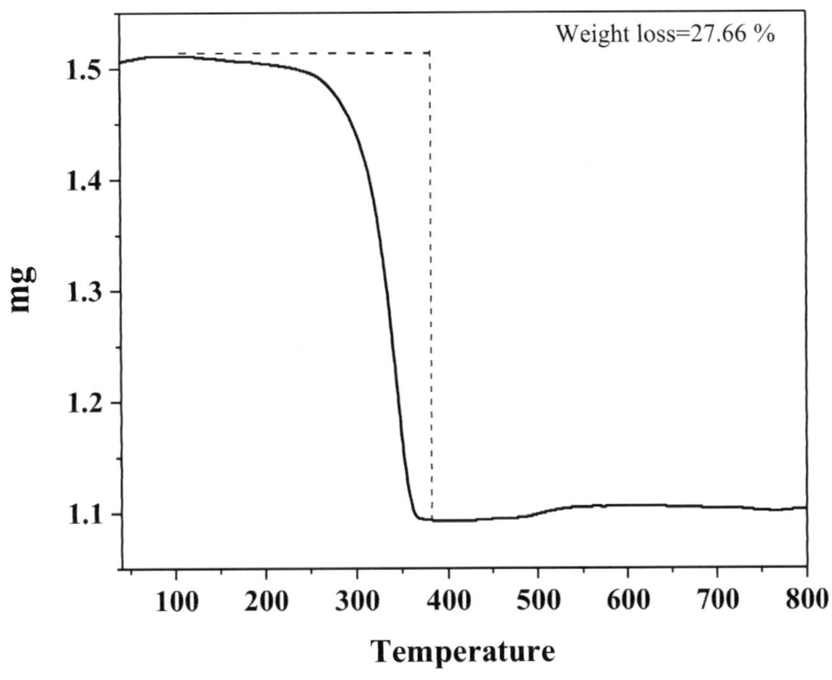

Figure 4. TGA curve of $Cd(OH)_2$ microstructures.

Thermogravimetric (TGA) analyses was approved for as grown **β**-$Cd(OH)_2$ as displayed in Fig.4. Data were plotted and analyzed using the first derivative of the mass loss versus temperature in °C. The TGA curve designates $Cd(OH)_2$ decomposition through the weight loss stages. The loss of weight could be attributed mainly to adsorption of molecules of water on surfaces of $Cd(OH)_2$ [**Error! Reference source not found.**]. This decomposition of $Cd(OH)_2$ to CdO micostructures has performed at 375 °C where the wight losses from 1.511 to 1.093 mg by percent 27.66 %. This percentage is twice for water loss (14 %) as calculated from decomposition equation but this could explained as suggested before for H_2O accumulation on $Cd(OH)_2$ surfaces [**Error! Reference source not found.**]. TGA result confirmed that 400°C is the optimum temperature to obtain pure single phase of CdO microstructures.

$$Cd(OH)_2 \longrightarrow CdO + H_2O$$

4. Conclusion

In summary, single phase of CdO microstructures could be successfully fabricated employing annealing temperature at 400 °C where β-$Cd(OH)_2$ thermal decomposition take place with percent 27.66 % from the total wight. Calcination of β-$Cd(OH)_2$ improve the particle size. TEM analysis of CdO displays that quasi-spherical and nanorice structure with size of about 151 and 173 nm was obtained, respectively. The results confirmed the materialization of CdO cubic phase at 400 °C where this could functional to fabricate optoelectronic applications based on pure cubic phase CdO.

Acknowledgments

Funding and support by Academy of Scientific Research and Technology (ASRT) Project ID: **6690** is gratefully acknowledged.

References

[1] Abdelfatah, M., Ledig, J., El-Shaer, A., Wagner, A., Marin-Borras, V., Sharafeev, A., Lemmens,P., Mosaad, M.M., Waag, A., & Bakin, A. (2016). Solar Energy Materials and Solar Cells. **145,** 454-461.

[2] Abdelfatah,M., Ledig, J., El-Shaer, A., Wagner,A., Sharafeev, A., Lemmens, P. , Mosaad, M.M., Waag, A., & Bakin, A.(2015). Solar Energy. **122,** 1193-1198.

[3] Abdelfatah, M., . Ledig, J., El-Shaer, A. , Sharafeev, A., Lemmens, P., Mosaad, M.M., Waag, A., & Bakin, A. (2016). ECS Journal of Solid State Science and Technology, **5,** 183-187.

[4] Abdelfatah, M., Ismail,W., & El-Shaer, A.(2018)Materials Science in Semiconductor Processing, **81,** 44-47.

[5] El-Shaer, A., Ismail,W. , & Abdelfatah,M. (2019). Materials Research Bulletin,**116,** 111-116.

[6] El-Shaer, A., Bakin, A., Al-Suleiman, M., Ivanov, S., Mofor, A.C., & Waag, A. (2007). Superlattices and Microstructures, **42,** 129-133.

[7] El-Shaer, A., Abdelfatah,M., Basuni, A., & Mosaad, M. (2018). Chinese journal of physics, **56,** 1001-1009.

[8] El-Shaer, A., Abdelwahed, A.R., & Mosaad, M. (2015). International Journal of Research in Engineering and Science, **3,** 28-36.

[9] Abdelfatah,M., Ismail,W., El-Shafai, N.M., & El-Shaer, A. (2020). Materials Technology, 1-9.

[10] Prakash, T., Neri, G., & Kumar, E. R. (2015). *Journal of Alloys and Compounds*, ***624,*** 258-265.

[11] Ferro, R., & Rodrıguez, J. A. (2000). Solar energy materials and solar cells, **64** (4), 363-370.

[12] Sravani, C., Reddy, K.T.R., Hussain, O.M., & Reddy, P.J. (1996). J Solar. Energy. Soc. India, **1** , 6-11.

[13] SU, L. M., GROTE, N., & SCHMITT, F. (1984). *Electronics Letters*, **20** (18), 716-717.

[14] Benko, F.A., & Koffyberg, F.P. (1986). Solid State Commun, **57 (12),** 901-903.

[15] Kondo, R., Okimura, H., Sakai, Y., & Jpn. (1971). J Appl. Phys, **10,** 1547.

[16] Shiori, A., & Jpn.(1997). Patent No., **7,** 909.

[17] Tadjarodi, A., Imani, M., Kerdari, H., Bijanzad, K., Khaledi, D., & Rad, M. (2014). Nanomaterials and Nanotechnology, **4,** 16.

[18] Askarinejad, A., & Morsali, A. (2009). Chem. Eng. J,**150,** 569–571.

[19] Choi, D., Jeong, G.H., Kim, S., & Bull,W. (2011). Korean Chem. Soc, **32,** 3851–3852.

[20] Bunge, S.D., Krueger, K.M., Boyle, T.J., Rodriguez, M.A., Headley, T.J., & Colvin ,V.L. (2003). J. Mater. Chem, **13,** 1705–1709.

[21] Mohamed, S. H., Hadia, N.M.A., Diab, A.K., & Abdel Hakeem, A.M. (2014). J. Alloys Compd., **609**, 68-72.

[22] Srinivasaraghavan, R., Chandiramouli, R., Jeyaprakash, B.G., & Seshadri, S. (2013). Spectrochimica Acta Part A: Molecular and Biomolecular Spectroscopy, **102**, 242-249.

[23] Kim, J.H., Hong, Y.C., Uhm, H.S., & Jpn. (2007). J. Appl. Phys, **46,** 4351.

[24] Zhang, Y.C., & Wang, G.L. (2008). Materials Letters, **62**(4-5), 673-675.

[25] Wang, W. W., & Zhu, Y. J. (2004). Inorganic Chemistry Communications, 7(9), 1003-1005.

[26] Athar, T., Shafi, S. S., & Khan, A. A. (2014). Materials Focus, *3*(5), 397-400.

[27] Saghatforoush, L. A., Sanati, S., Mehdizadeh, R., & Hasanzadeh, M. (2012). Superlattices and Microstructures, **52,** 885-893.

[28] Shanmugam, N., Saravanan, B., Reagan, R., Kannadasan, N., Sathishkumar, K., & Cholan, S. (2014). Moern Chemistry & Applications, **2.**

[29] Usharani, K., Balu, A.R., Nagarethinam, V.S., & Suganya, M.(2015). Prog. Nat. Sci. Mater. Int, **25** (3), 251- 257.

[30] Manjula, N., Puganenthi, M., Nagarethinam, V.S., Usharani, K., & Balu, A.R. (2015). Mater. Sci. Pol., **33** (4), 774- 787.

[31] Javan, N. S., Azizian-Kalandaraghm, Y., & Khodayari, A. (2013). Optoelectronics and Advanced Materials-Rapid Communications, **7** (9-10), 639-642.

[32] Mohanraj, K., Balasubramanian, D., Jhansi, N., Bakkiyaraj, R., & Chandrasekaran, J. (2017). Int. J. Thin. Fil. Sci. Tec., ***6***, 87-91.

[33] Moorthy, A., Subramaniam, M. R., Manivasagam, T. G., & Kumaresan, D. (2018). Dalton Transactions, 47(26), 8683-8689.

IOP Conf. Series: Materials Science and Engineering **956** (2020) 012007 doi:10.1088/1757-899X/956/1/012007

The Effect of burnable absorbers on neutronic parameters of VVER-1200 reactor

Noura Hafez[1], Hesham Shahbunder [2,3], Esmat Amin [4], S. U. El-Kamessy [3], S. A. Elfiki [3] and Ahmed Latef [5]

[1] Nuclear Power Stations Engineering Department, Faculty of Engineering, Egyptian Russian University, Cairo, Egypt.
[2] Department of Physics, College of Sciences and Humanities in Al-Kharj, Prince Sattam bin Abdulaziz University, Al-Kharj 11942, Saudi Arabia.
[3] Physics Department, Faculty of Science, Ain Shams University, Cairo 11566, Egypt.
[4] Nuclear and Radiation Regulatory Authority (NRRA), Cairo, Egypt.
[5] Faculty of Engineering, Fayoum University, Cairo, Egypt.

E-mail: noura.physics@hotmail.com

Abstract. The presence of burnable absorbers play an important role in the safety of nuclear reactors. So, seeking their influences on the reactor core behaviour is a crucial issue in both reactor core design and operation. Therefore, the aim of the present work is to study the effect of the gadolinium oxide burnable absorbers on one of the significant safety parameters of VVER-1200/V392M core which is the neutron flux distribution. A full-scale three-dimensional model of the VVER-1200/V392M core was performed using MCNP5/MCNPX transport code with ENDF/B-VII.0 nuclear data library. The calculations were performed in the normal operation state for the first fuel cycle. Besides the neutron flux in both radial and axial distributions, the comparison between the fuel assemblies contained gadolinium oxide and that without gadolinium oxide is also discussed. Also, the effect of gadolinium oxide on the produced major actinides is investigated.

1. Introduction

The specification of the reactor core, that depends on its nuclear characteristics, is an essential aspect in nuclear safety and economic evaluation. These nuclear characteristics influence on the neutron behaviour and hence, the power distribution and fuel cycle which have to be extended to the improvement of fuel economics [1]. The improvement of fuel economics lead to increase fuel enrichment and also the need to improve the neutron flux distribution and reactivity control. These improvements in neutron flux distribution required the integration of the burnable absorbers in the fuel matrix. Burable absorbers (BA) or burnable poisons (BP) are materials such as gadolinium, dysprosium, boron, erbium and hafnium, which in their non-radiated state have a high neutron absorption cross-section. These materials are converted into materials of relatively low absorption neutron cross-section as a result of neutron absorption.

Analysis studies on the performance of different burnable absorbers were carried out which mostly focused on the boiling water reactors [2, 3, 4]. Fewer studies had faced the pressurized water reactors

[5, 6, 7]. Consequently, the present study is focused on studying the performance of burnable absorbers on PWRs especially that have hexagonal geometries.

The aim of the present work is studying and analyzing the influence of gadolinium oxide on the neutronic safety parameters such as neutron flux distribution including both fast and thermal distributions for the full core. Moreover, to investigate the effect of gadolinium oxide on the produced actinides in particularly the major actinides. Despite the importance of power distribution calculations among the whole core, the present work has focused on the neutron flux distributions to overcome the difficulty of power distributions due to the different materials utilized in the core.

The analysis of the present study is important not only for the safety of reactor operation, but also for enhancement of reactor core design.

2. VVER-1200 core description

In the present work, the VVER-1200 reactor core is selected as a case study that is one of the pressurized water reactors in which the borated water is utilized as coolant and moderator. The core is composed of hexagonal 163 fuel assemblies which are located on 13 hexagonal grids.
Each fuel assembly contains 331 rods that classified into three groups of 312 fuel rods, 1 instrumentation channel, and 18 guide channels in which the control rods are inserted. There are two types of fuel assemblies into the core corresponding to two different types of the fuel elements, which are uranium dioxide UO2 and mixture of uranium dioxide and gadolinium oxide $UO_2 + Gd_2O_3$ as shown in Fig. 1. Fig. 2 shows a horizontal cross-section MCNP5 model of VVER-1200 core including the different types of fuel assemblies with different enrichment of fuel elements and their distributions around the core. A detailed parameters of the VVER-1200 core geometry and materials used are described in Table 1[8, 9].

The regulation of reactor power and suppression of fission chain reaction is carried out by two systems adjusting reactivity. The first one is the control and protection system (cps), in which 121 control rods divided into 12 groups are used for changing reactivity and for reactor shutdown in normal and emergency operation conditions. The second is the boron regulation system that is used for small changes in reactivity by changing boric acid concentration in the water primary circuit [8, 10].

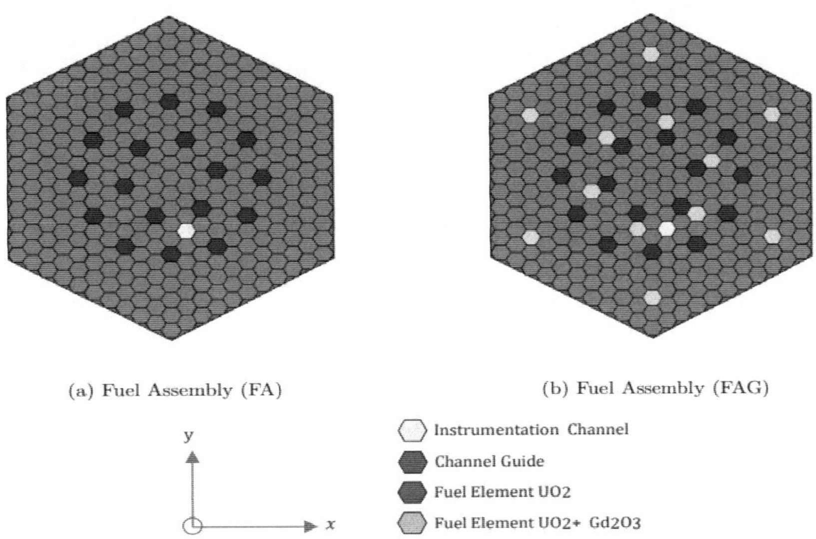

Fig. 1. Radial view of the two fuel assembly types.

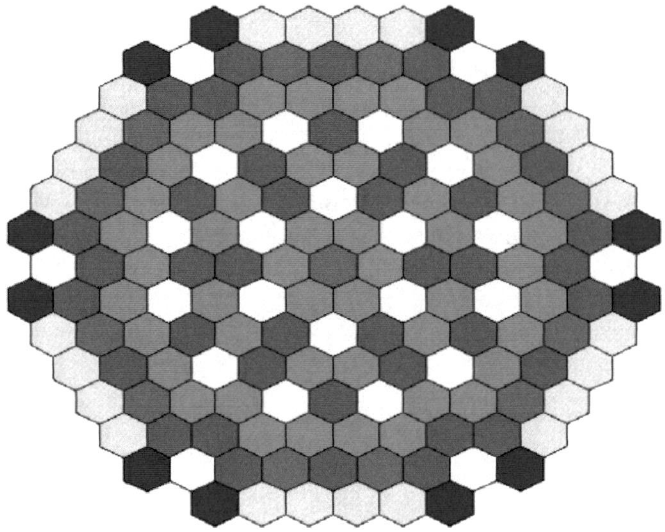

FA-Z13 [UO_2 - 1.3 wt.% U-235]
FA-Z24, [UO2 - 2.4 wt.% U-235]
FA-Z40 [UO_2 – 4.0 wt.% U-235]
FAG-Z33Z9 [303 FE(UO_2) - 3.3 wt.% U-235] + [9 FE (UO_2 + Gd_2O_3) – 2.4 wt.% U-235 & 8wt.% Gd_2O_3]
FAG-Z44B2 [300 FE(UO_2) - 4.4 wt.% U-235] + [12 FE (UO_2 + Gd_2O_3) – 3.6 wt.% U-235 & 5wt.% Gd_2O_3]
FAG-Z33Z2 [300 FE(UO_2) - 3.3 wt.% U-235] + [12 FE (UO_2 + Gd_2O_3) – 2.4 wt.% U-235 & 8wt.% Gd_2O_3]

Fig. 2. The radial view of the VVER-1200 reactor core.

3. Computational tools

In the present study, the three dimensional model of VVER-1200 core were executed by Monte Carlo N Transport Code MCNP5 [11, 12] with ENDF/B-VII.0 nuclear data library [13] to calculate the neutronic safety parameters of VVER-1200 including the effective multiplication factor k_{eff}, the total, thermal and fast neutron flux for both axial and radial distributions, in addition to the burn-up calculations that were executed using MCNPX 2.7.0 code [14] with the same nuclear data library. The total number of histories employed in the eigenvalue calculations was 35×10^6 including the tracking of 100000 neutrons per time step with 250 active cycles and the standard deviation was less than 0.00015 in one σ, for all calculations.

The total integrated flux was typically calculated using FMESH:n superimposed mesh tally card with EMESH energy card to identify the thermal and fast energy regions. Since the obtained results from MCNP are normalized to one source neutron, so the obtained results from MCNP5 have to be scaled to the steady-state thermal power level of the critical state system as follows.

$$\varphi_{normalized}\left(\frac{neutrons}{cm^2 . s}\right) = \nu\left(\frac{neutrons}{fission}\right) \times \frac{3200 \times 10^6 (W)}{1.6022 \times 10^{-13}(\frac{J}{MeV}) \times \epsilon_f(\frac{MeV}{fission})} \times \varphi_{MCNP}\,(\frac{1}{cm^2})$$

IOP Conf. Series: Materials Science and Engineering **956** (2020) 012007 doi:10.1088/1757-899X/956/1/012007

Table 1

The VVER-1200 core description [8],[9]

Parameter	Value
Thermal power	3200 MW
Electrical power	1200 MW
Coolant, Moderator	Purified water + boric acid
Boric acid concentration in water	1.0-16.0 g - BH_3O_3/Kg - H_2O
Fuel Assembly (FA)	
FA "Turnkey" size	
Maximum	23.51 cm
Nominal	23.42 cm
FA height	457 cm
Gap between fuel assemblies	34 cm
Fuel Element (FE)	
Fuel Type	UO_2
	$UO_2 + Gd_2O_3$
Enrichment	1.3 to 4.5 wt% U-235
Mass fraction of Gd2O3 in fuel	5.0 to 8.0 wt% Gd_2O_3
Shape of FE	Rod
Fuel height	373 cm
Fuel pellet diameter	0.76 cm
Central hole diameter	0.12 cm
Clad material	E-110 alloy
Clad density	6.515 g/cm^3
Clad height	393 cm
Outer diameter of clad	0.91 cm
Inner diameter of clad	0.773 cm
Control rod	
Outer diameter of guide channel	1.29 cm
Inner diameter of guide channel	1.09 cm
Clad material	SS alloy 12H18N10T
Clad density	7.85 g/cm^3
Absorber materials	Boron Carbide (B_4C)
	Dysprosium titanate ($Dy_2O_3TiO_2$)
Absorber materials density	1.8 g/cm^3 (B_4C)
	4.85 g/cm^3($Dy_2O_3TiO_2$)
Instrumentation channel (IC)	
Outer diameter of IC	1.29 cm
Inner diameter of IC	1.09 cm
Clad material	Zr alloy E-635
Spacer grid	
Spacer grid material	SS 12H18N10T
Spacer grid density	7.85 g/cm^3

IOP Conf. Series: Materials Science and Engineering **956** (2020) 012007 doi:10.1088/1757-899X/956/1/012007

4. Results and discussion

4.1. Mode of operation

The neutronic parameters calculations of VVER-1200 reactor core were accomplished for the normal operation state, where the criticality (k_{eff} = 1.00018 ± 0.00012) achieved under the following conditions shown in Table 2.

Table 2
Normal operation state conditions.

Parameter	Value
Average coolant temperature	586 K
Average fuel temperature	1073 K
Boric acid concentration	7.44 g/kg – H_2O
Inserted control rods	Group no. 12 (inserted by 20 %)

4.2. Neutron flux distribution

4.2.1. Radial neutron flux distribution.

The radial total, thermal and fast flux distributions along *x*-axis for VVER-1200 core in normal operation state in case of the presence of gadolinium rods are shown in Fig. 3. As shown in Fig. 3, the total flux distribution along *x*-axis has a central down curvature in the region from $x = -50$ cm to $x = 50$ cm, in addition to slightly depression in sides regions from $x = -100$ cm to $x = -75$ cm and from $x = 75$ cm to $x = 100$ cm. By comparing the total, thermal and fast neutron flux distributions, one can observe that both the thermal and fast neutron flux distributions have central down curvature in the same region as in the total distribution and only the fast neutron distribution has the slight decrease in the same sides regions as the total flux distribution. This is due to the contribution of the fuel assemblies (FAG) that contain to the gadolinium oxide (gadolinia) rods, where 24 FAGs of the type FAG-Z33Z9 and 13 FAGs of type FAG-Z33Z2 are distributed uni-formally from the center of the core and in the core corner. Besides 24 FAGs of type Z44B2 that are located in the outside border of the hexagonal core.

To study the effect of gadolinium rods on the neutron flux distribution, the VVER-1200 core was modeled without any gadolinium rods (clean core). The comparison between the radial total, thermal and fast neutron flux distributions along *x*-axis for both cases without and with gadolinium rods is shown in Fig. 4, Fig. 5 and Fig. 6 respectively. In case of the core without Gd_2O_3 rods, both the total, fast neutron flux distributions have a very slight depression in the core center from $x = -5$ cm to $x = 5$ cm which doesn't appear in the thermal flux distribution. In addition to all the neutron flux distributions have the two down curvature at $x = -75$ cm and $x = 75$ cm.

From the comparison between all neutron flux distributions in both cases with and without Gd_2O_3 rods, it is clear that the gadolinium is a good neutron burnable poison. this is due to the high absorption cross-section at thermal energy especially for the two isotopes ^{157}Gd and ^{155}Gd as shown in Fig. 7. Moreover, it forms daughter products as a result of neutron absorption as Sm and Eu isotopes with relatively low absorption cross-sections of 5670 b and 4565 b respectively. The central slight depression in the region from $x = -5$ cm to $x = 5$ cm in case of the core without gadolinia and also the two down curvature in the sides regions in both cases of core are formed due to the insertion of the control rods of group 12 by 20 %, where the inserted part is made of Dysprosium titanate ($Dy_2O_3TiO_2$) that has absorption cross-section at thermal energy of 940 b.

IOP Conf. Series: Materials Science and Engineering **956** (2020) 012007 doi:10.1088/1757-899X/956/1/012007

Due to the symmetry of the core, the radial total, fast and thermal flux distributions along y-axis are approximately similar to those along x-axis.

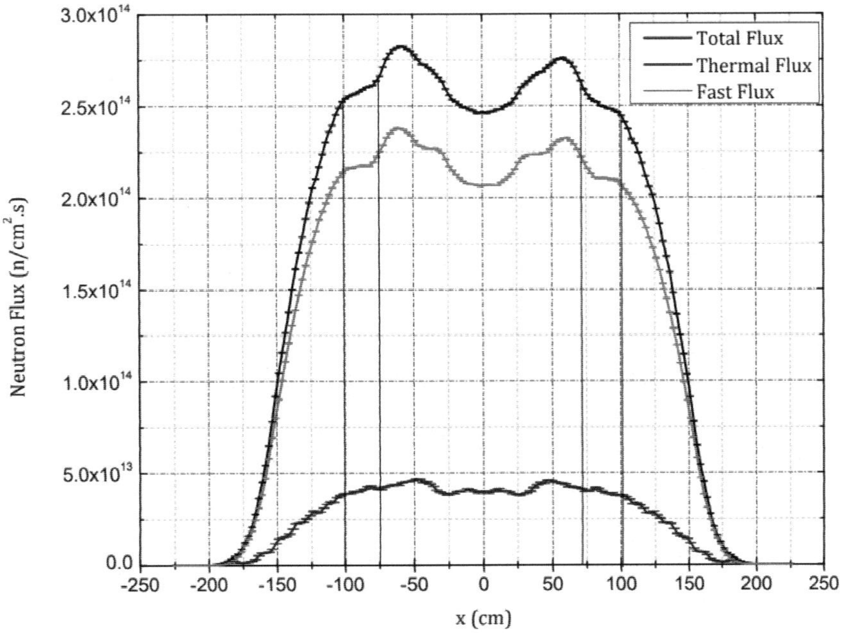

Fig. 3. the radial total, thermal and fast flux distributions along x-axis of VVER-1200 core

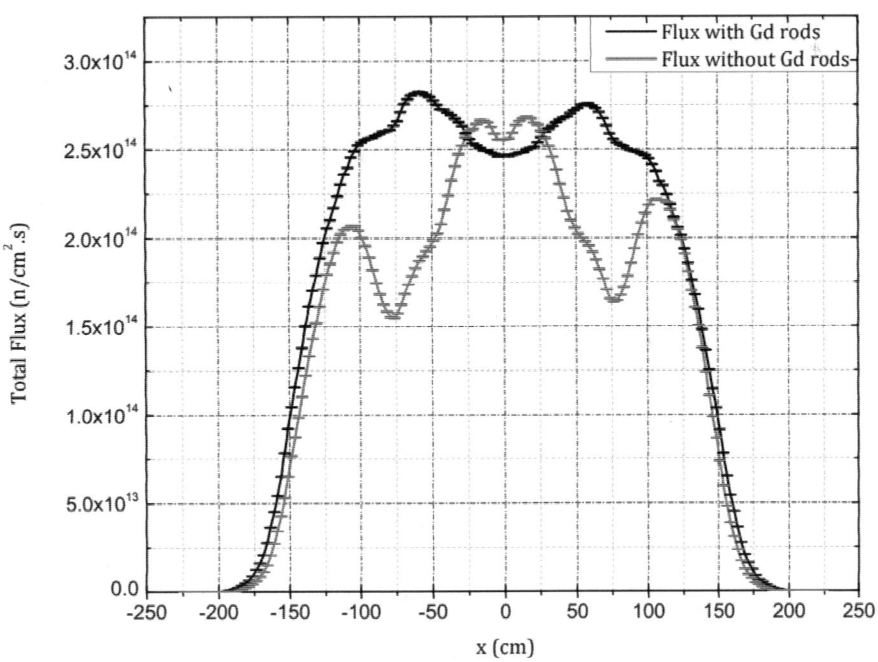

Fig. 4. The comparison between total radial flux distributions along x-axis in case of with and without Gd rods

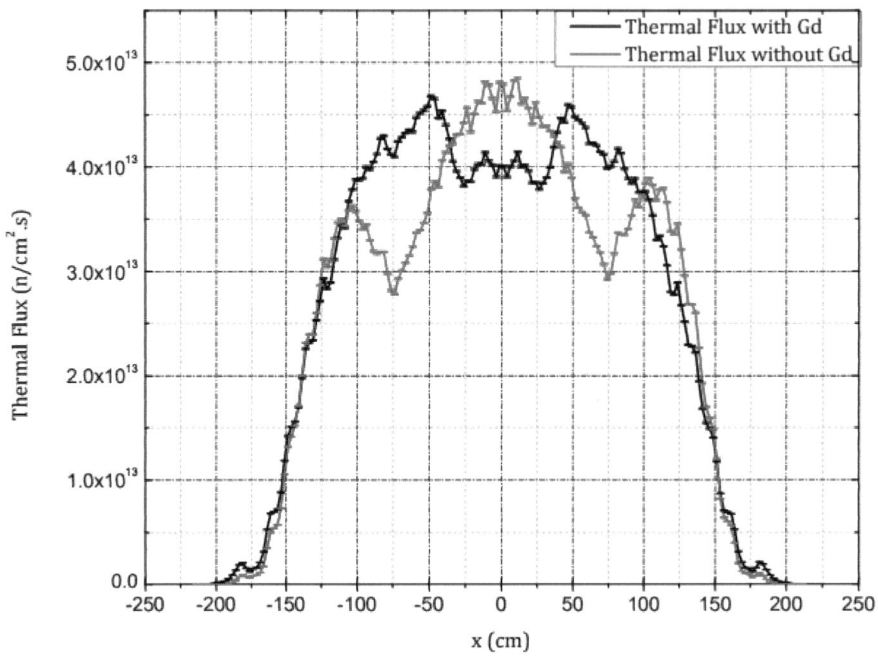

Fig. 5. The comparison between thermal radial flux distributions along x-axis in case of with and without Gd rods

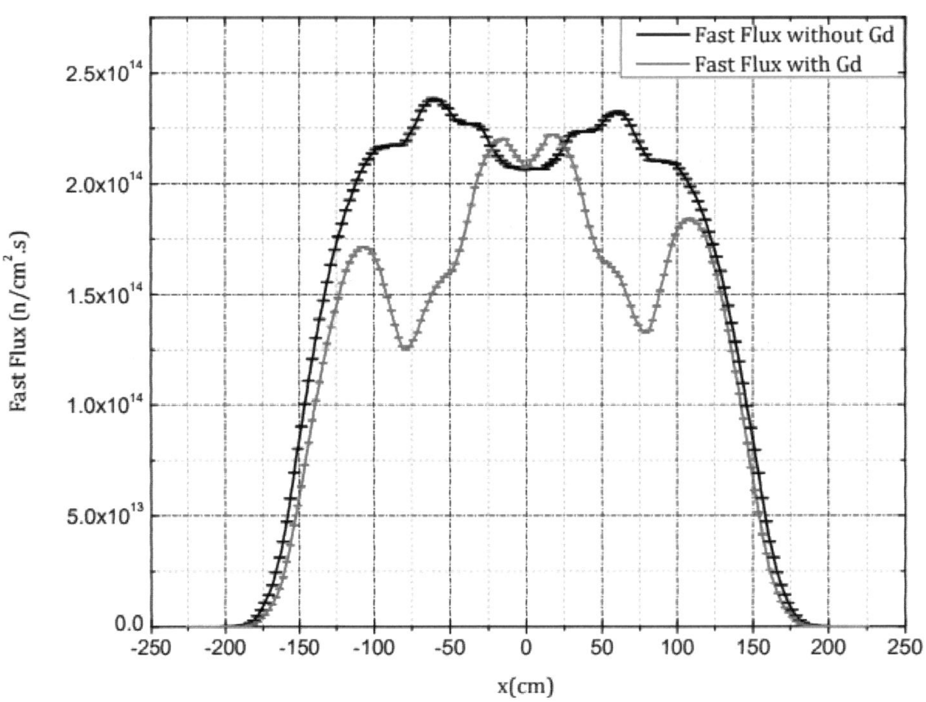

Fig. 6. The comparison between fast radial flux distributions along x-axis in case of with and without Gd rods

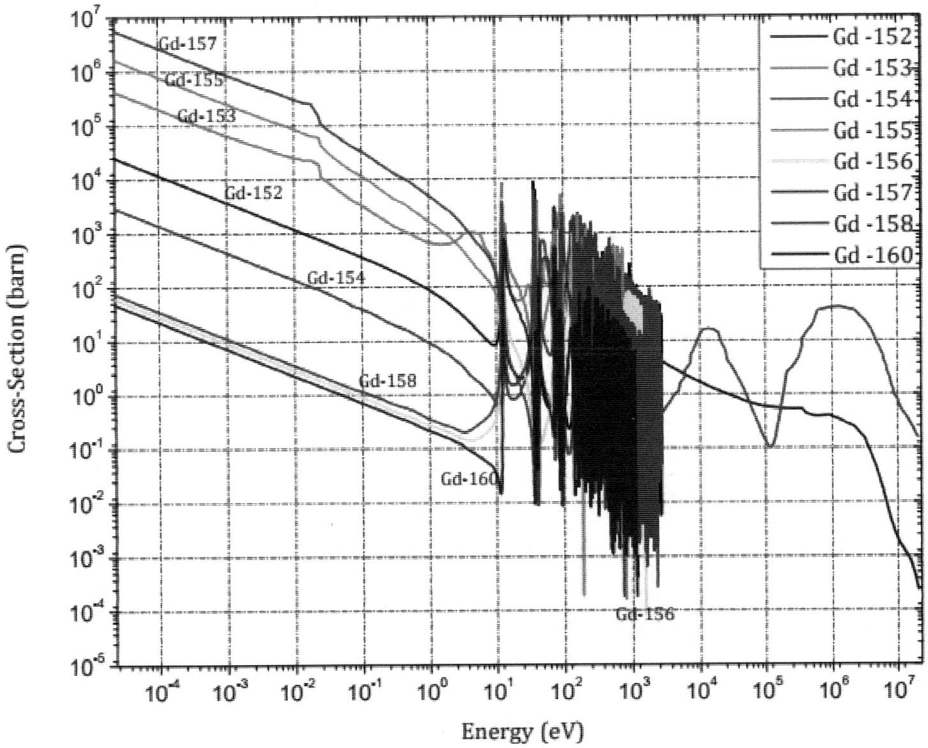

Fig. 7. The comparison between cross-sections of the Gd isotopes for (n,γ) reaction

4.2.2. Axial neutron flux distribution.

In case of presence of gadolinium rods, the axial neutron flux for the whole VVER-1200 core, including the total, thermal and fast distributions, is shown in Fig. 8. From Fig. 8, one can observe that all distributions are generally symmetric around the center of the core. This is as a result of the symmetry in fuel distributions among the whole core. But by comparison between total, thermal and fast neutron distributions for the core without any gadolinium rods and with the presence of gadolinium rods as shown in Fig. 9, Fig. 10 and Fig. 11. respectively, it was noticed that despite the central down curvature that observed in radial distribution, all distributions (total, thermal, and fast) are approximately the same.
The calculations of axial and radial power distributions along the whole are complicated processes because the power distribution results from the integration of several physical parameters related to the flux distribution. So the neutron flux distribution is considered as an indicator for the power distribution along the core that is one of the important safety operation parameters.

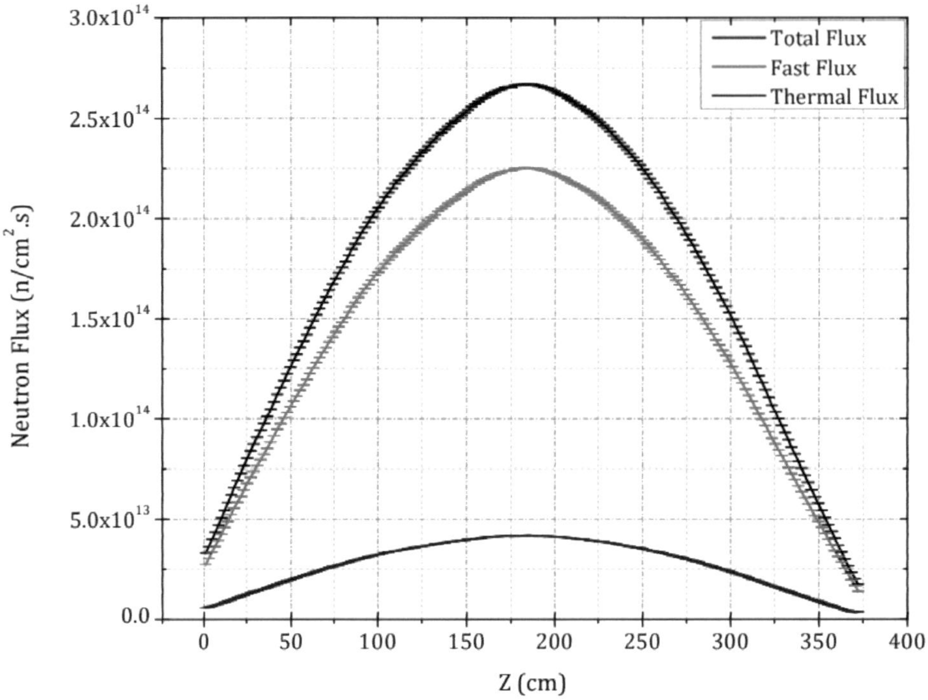

Fig. 8. The axial total, thermal and fast flux distribution of VVER-1200 core

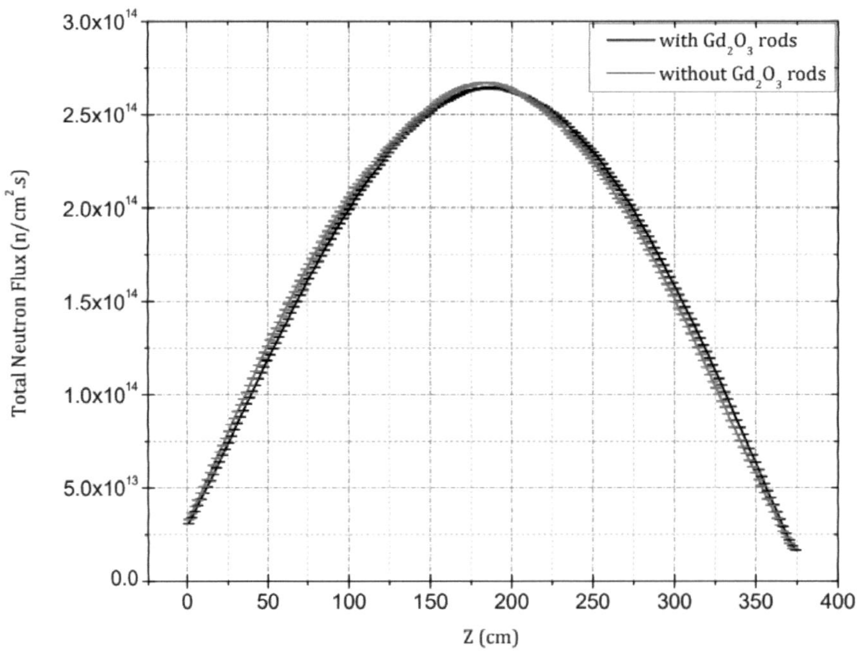

Fig. 9. The comparison between total axial flux distributions along z-axis in case of with and without Gd rods

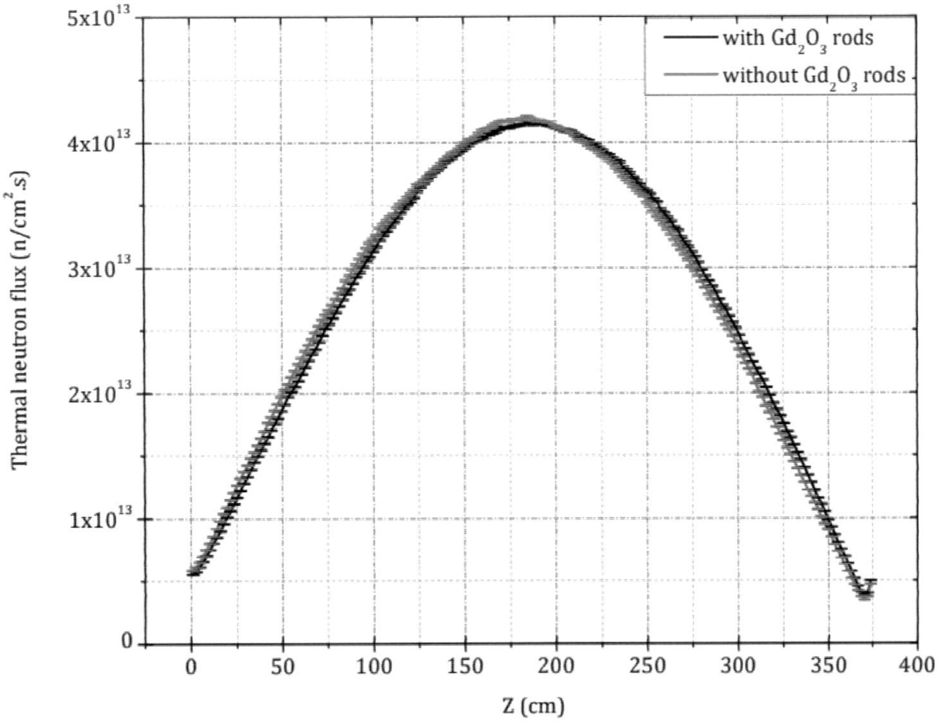

Fig. 10. The comparison between thermal axial flux distributions along z-axis in case of with and without Gd rods

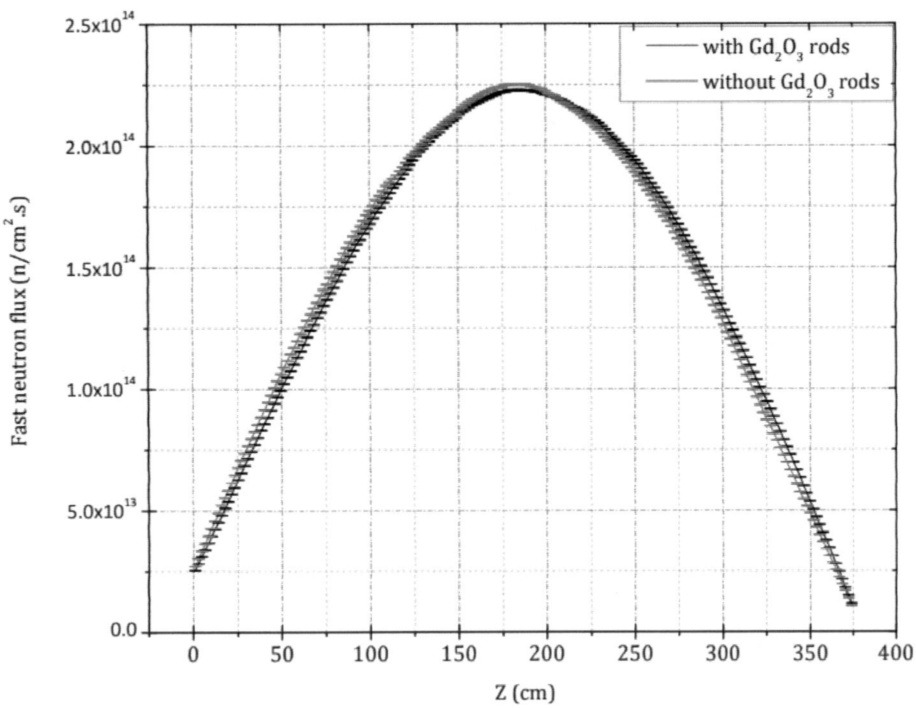

Fig. 11. The comparison between fast axial flux distributions along z-axis in case of with and without Gd rods

4.3. Burnup

4.3.1. Actinide isotopes.

The produced major actinides including plutonium and uranium isotopes from the three different types of fuel assemblies, where one doesn't contain gadolinium and the others containing gadolinium with different numbers of rods, after burning of 600 days operation cycle are presented in Table 3. By comparing the mass variation of produced actinides among the different fuel assembly types, one can observe that the FAs don't containing gadolinium fuel rods yield large amount of actinides than the second type containing gadolinium fuel rods except ^{239}Pu which produced in larger amount from FAs containing gadolinium fuel rod This is as a result of the presence of gadolinium leads to increasing the absorption probability of fast neutrons by ^{238}U, hence increasing the amount of ^{239}Pu.

The analysis of plutonium isotopes are essential where they contribute the fission reaction process and determine the spent fuel proliferation level.

Table 3
Mass of actinide isotopes.

Isotope	Mass (g)		
	Z24	Z33Z9	Z33Z2
^{235}U	3.1×10^3	5.8×10^3	5.8×10^3
^{238}U	4.4×10^5	4.4×10^5	4.4×10^5
^{238}Pu	3.4×10^1	3.2×10^1	3.3×10^1
^{239}Pu	2.9×10^3	3.2×10^3	3.2×10^3
^{240}Pu	8.9×10^2	7.8×10^2	7.8×10^2
^{241}Pu	5.6×10^2	5.3×10^2	5.4×10^2
^{242}Pu	1.6×10^2	1.1×10^2	1.1×10^2

5. Conclusion

The effect of burnable absorbers including gadolinium and dysprosium titanate in the first cycle of the VVER-1200 reactor was studied using MCNP5/MCNPX codes.

The presence of gadolinium oxide plays a significant role in the safety of reactor, where it makes the neutron flux much flattened, and hence minimizes the power peaking specially at the canter of the core. Also, dysprosium titanate is found to be good control rods where it increases the safety of operation by the relatively high absorption cross-section.
From the present study, it was noticed that the amount of produced plutonium isotopes, except ^{239}Pu, from the fuel assemblies that don't contain gadolinium oxide, are higher than that of contain gadolinium. This is because of the shielding effect of fissile isotopes as a result of presence of gadolinium oxide. The analysis of produced actinides in particularly the major actinides is important in the reactor operation as it determines the spent fuel proliferation level.

Further study of other burnable poisons in different core configurations with additional analysis is required to obtain the optimum reactor core designs.

Acknowledgment
The authors would like to gratefully thank Rosatom Technical Academy [15] members for providing some required information for this work.

References

[1] IAEA 1996 Safe core management with burnable absorbers in WWERs Tech. Rep. IAEA-TECDOC-858 Nuclear Power Technology Development Section Vienna, Austria

[2] Tung W H and Lee T T 2020 Annals of Nuclear Energy 138 107204 URL https://doi.org/10.1016%2Fj.anucene.2019.107204

[3] Radaideh M I, Price D, O'Grady D and Kozlowski T 2019 Progress in Nuclear Energy 113 230–246 URL https://doi.org/10.1016%2Fj.pnucene.2019.01.010

[4] Galahom A, IIBashter and Moustafa Aziz 2015 Annals of Nuclear Energy 76 461–468 URL https://doi.org/10.1016%2Fj.anucene.2014.10.025

[5] Mustafa S and Amin E 2020 Radiation Physics and Chemistry 171 108724 URL https://doi.org/10.1016%2Fj.radphyschem.2020.108724

[6] Tran H, Hung Hoang and Peng Hong Liem 2017 Energy Procedia 131 29–36 URL https://doi.org/10.1016%2Fj.egypro.2017.09.442

[7] Galahom A A 2016 Annals of Nuclear Energy 94 22–31 URL https://doi.org/10.1016%2Fj.anucene.2016.02.025

[8] Dien L D and Diep D N 2017 World Journal of Engineering and Technology 05 507–519 URL https://doi.org/10.4236/wjet.2017.53043

[9] IAEA 1995 In-core fuel management code package validation for WWERs Tech. Report IAEA-TECDOC-847 International Atomic Energy Agency Vienna URL https://inis.iaea.org/search/search.aspx?orig_q=RN:28026025

[10] IAEA 2003 WWER-1000 reactor simulator (Training Course Series no 21) (Vienna: International Atomic Energy Agency) URL https://www.iaea.org/publications/6686/wwer-1000-reactor-simulator

[11] Brown F, Kiedrowski B and Bull J 2010 MCNP5-1.60 release notes Tech. Report LA-UR-10-06235 LosAlamos National Laboratory (LANL) New Mexico, U.S. URL https://mcnp.lanl.gov/pdf_files/la-ur-10-06235.pdf

[12] X5 Monte Carlo Team 2003 MCNP – A general Monte Carlo N-Particle Transport code, version 5 Tech.Report LA-CP-03-0245 Los Alamos National Laboratory (LANL) New Mexico, U.S. (Revised 2/1/2008) URL https://mcnp.lanl.gov/pdf_files/la-cp-03-0245.pdf

[13] Chadwick M B, et. al. 2006 Nuclear Data Sheets 107 2931–3060 URL http://dx.doi.org/10.1016/j.nds.2006.11.001

[14] Pelowitz B 2011 MCNPX User's Manual Tech. Rep. LA-CP-11-00438 Los Alamos National Laboratory (LANL)

[15] Rosatom 2020 Rosatom Technical Academy Ulitsa Kurchatova, 21, Obninsk, Kaluga Oblast, Russia, 249031(Accessed: January 13, 2020) URL http://rosatomtech.com

IOP Conf. Series: Materials Science and Engineering **956** (2020) 012008 doi:10.1088/1757-899X/956/1/012008

Attenuation Effectiveness of Double Phase Stainless Steel Alloys for Fusion Reactor System

Noha M. Ali[1], Aly Saeed[1], R. M. El Shazly[2], S. A. Al-Fiki[3], M. M. Eissa[4], S. U. El-kameesy[3]

[1]Nuclear Power Stations Department, Faculty of Engineering, Egyptian-Russian University, Cairo, Egypt. [2]Physics Department, Faculty of Science, El-Azhar University, Cairo, Egypt. [3] Physics Department, Faculty of Science, Ain Shams University, Cairo, Egypt. [4] Steel Technology Department, Central Metallurgical Research and Development Institute, Helwan, Egypt.

E-mail: Noha.physics@yahoo.com.

Abstract. Aluminium, tungsten and aluminium – tungsten ferritic stainless-steel alloys were developed and produced using 30 kg pilot plant induction furnace of medium frequency at the same conditions to be applied in nuclear reactor shielding material. Higher hardness, ultimate strength, and ultimate tensile strength are appeared in the developed alloys. Aluminium stainless-steel elongation had lower values. Total slow neutrons, neutrons > 10 keV, and slow neutrons of aluminium-tungsten based ferritic stainless steel alloys which carried out using ^{241}Am-Be neutron source as well as a wide range up to 1400 keV energies of gamma rays, which emitted from Co-60, Cs-137 and Eu-152, were used by means of sodium iodide (Tl) detector and studied using the XCOM program. Results showed that all studied stainless-steel alloys own a converge values of neutron macroscopic cross sections (Σ, cm^{-1}). Moreover, there is no significant variation of the value of mass attenuation coefficients (σ, cm^2/g) for the studied alloys. Good agreement between both experimental and calculated results of mass attenuation coefficients was achieved. A new composition for several nuclear applications such as, nuclear shielding applications were achieved.

1. Introduction

Energy from fusion system own properties of an appealing product with respects to environmental advantages and safety compared with other energy sources [1,2]. Moreover, fusion fuels are obtainable in the nature, opposite to rare fission fuel resources. Then, in the past 40 years many studies concentrated on fusion energy discussion [3].

Fusion reactors have been predictable to work at high temperatures and utilize chemically aggressive coolants such a normal lithium. And therefore, much kinds of materials are needed as covering materials in the fusion reactor. In addition, fusion reactor first walls around the fusion cell must withstand gamma-ray radiation and high energetic strong neutron fluxes, whose are predictable to command to much higher material harm than observed by fission reactors [4,5].

So, chosen of appropriate structural material plays a keys function in designing fusion reactors [6,7]. Stainless steel is such grade of steel that can be applied in severe environmental conditions. In high-temperature medium, such as that of nuclear reactors, suitable radiation resistance materials are needed [8-12]. Stainless steel alloys were considered to be utilized extensively for structural components in more reactor system designs and in different high temperature applications [8]. Due its special mechanical property such, formability and toughness, large surface area, high strength, and excellent thermal and physical properties, stainless steels be proper as nuclear reactor materials [13,14].

Ferritic stainless steel has the lion share in applications that are required for structure stability at high temperature, and corrosion resistance as well [15,16]. Good formability with competitive low-cost ferritic stainless steels, oxidation resistance and excellent corrosion are favorable materials to change austenitic stainless steels, for using in solid oxide fuel cells, automobile exhaust systems, architecture applications and petroleum refining plants [17-19]. Due to the high resist for corrosion and thermal properties, good mechanical strength and formability, and low cost, ferritic stainless steel alloys are widely used for high temperature applications [14]. Precipitation of chromium carbides, other intermetallic compounds based on chromium and grain growth of ferrite at high temperature are considered the most challenges, facing the application of ferritic stainless steel alloys(FSSs) in nuclear engineering material [16]. By Increasing the property of gamma attenuation of ferritic stainless-steel alloys will raise its multifunction using for more nuclear reactor implementation [20].

At the meantime, aluminum has been proved among the most ferrite stabilizer alloying elements, adding to its capability for promoting the generation of equ-axis ferrite structure [21].

Tungsten is recently proposed for inhibiting the affinity of chromium in forming coarse laves, substituting it by more fine intermetallic precipitates. To improve the mechanical properties of steels, tungsten (W) is generally added as strong ferrite former. Increasing the steel yield strength and hardenability, that was accompanied with decreasing charpy impact energy and elongation were observed in many steel alloys due to adding W. Tungsten alloys used for the helium-cooled diverter and the safeguard of the helium-cooled initial wall layer in reactor designing going beyond ITER has been investigated and discussed for several years [22-25].

The studied alloys have a mix of austenite, ferrite and martensitic in their structure. They exhibit characteristics of all phases with higher ductility and strength. Thereby, this work was designed to study the effect of partial modification of the chemical composition of ferritic stainless steel by adding carbon, tungsten, and aluminum to attain the most proper ferritic stainless steel required for serving at the conditions of different nuclear material applications such as, high-temperature segments in primary piping systems, radiation shielding domain and reactor pressure vessels.

Due to the previous purposes, developed study is concentrated on studying Al-W stainless steel. Then, the attenuation properties of the produced stainless steel alloys against neutrons and gamma ray were calculated and compared.

2. Experimental

2.1. Samples preparation

Four heats were melted in 30 kg magnesite crucible induction furnace through melting of ferritic stainless steel scrap. One heat was kept as reference, while the other three heats were modified by adding different amounts of carbon, tungsten, titanium, and aluminum. The chemical composition of the four heats were recorded in Table 1. The molten metal was casted into metal mold with dimensions 70*70*500 mm. Then, the ingots were forged at starting temperature 1050°C, and finished at 950°C to attain 30*30 mm square cross section rods. Samples were cut and prepared for different investigations. Solution treatment was applied on all samples by heating at 1100°C, directed by quick quenching in water.

Table 1. Chemical composition of studied stainless steel alloys.

	Alloy code			
Elements	FS3	FS8	FS11	FS12
C	0.053	0.441	0.071	0.084
Mn	0.633	0.892	0.754	0.744
Cr	13.800	12.400	13.800	13.900
Ni	0.079	0.078	0.101	0.113

Mo	0.002	0.008	0.021	0.036
Al	0.001	0.110	0.152	0.211
Co	0.019	0.020	0.011	0.009
Cu	0.077	0.073	0.086	0.142
Nb	0.016	0.027	0.036	0.046
Ti	0.003	0.003	0.064	0.403
V	0.059	0.068	0.055	0.057
W	0.005	0.005	1.610	2.840
Pb	0.005	0.005	0.005	0.005
Ca	0.001	0.008	-	-
N	0.055	0.028	0.038	0.035
Fe	Balance	Balance	Balance	Balance

2.2. Mechanical testing

Vickers hardness experiments were achieved on high polished surface of stainless steel samples. Zwick-Roel hardness machine with 50 kg working loading is utilized for hardness measurements. The average from many readings was possessed. Tensile test measured for samples at room temperature and yield strength, ultimate, elongation and impact energy were measured. Tensile specimens with dimensions accordingly ASTM-E8 specification. Each of the tensile values was the average of two tensile test data.

2.3. Gamma-rays attenuation measurements

Gamma rays emitted from 3.7μci Eu-152, 9.5 μci Cs-137, and 4.9μci Co-60 radioactive sources were applied as source of gamma ray energies. Eight gamma rays released tops (344.2, 661.6, 778.9, 96, 1112.4, 1173.2, 1332.5 and 1407.2 keV) of the power spectrum were selected to involve a broad range of energies to calculate the attenuation coefficients of the studied alloy barriers. Figure 1 displays the graph of gamma radiation detection technique, where a 3″ x 3″ sodium iodide (Tl) scintillation detector was applied to calculate the gamma ray intensity for the studied energy lines. The linear attenuation coefficient was evaluated by the Beer-Lambert's equation, consisting on the gamma-radiation concentration that transmitted during the investigated samples. Specimens were cut in a disc form and smooth to have parallel plates. The total gamma rays and half value layer (HVL) for each sample were obtained by the following equations [26].

$$\mu = \frac{\ln \frac{I_0}{I_x}}{x}, \qquad HVL = \frac{\ln 2}{\mu} \tag{1}$$

where μ (cm^{-1}) is called the linear attenuation coefficient, I_x and I_o are the concentration of gamma radiation after and before transmissions during the specimen, and x is called the width of sample.
Mass attenuation coefficient (σ, cm^2 /g) on the other hand, is a serious nuclear parameter independent of the bulk density of the substance, was evaluated by considering superficial density (xρ).

$$\sigma = \frac{\ln \frac{I_0}{I_x}}{x\rho} \tag{2}$$

The calculated values by "WinXCom" program (version 3.1) [27,28] for each sample composition based on the combined rule of the following equation were compared together with the experimental results of the mass attenuation coefficient:

$$\sigma_t = \sum_{i}^{n} W_i \left(\frac{\mu_i}{\rho_i}\right)_m \qquad (3)$$

Where $(\frac{\mu_i}{\rho_i})_m$ is the mass attenuation coefficient for every element in each mixture sample, while W_i is the partial weight of the elements within each mixture sample.

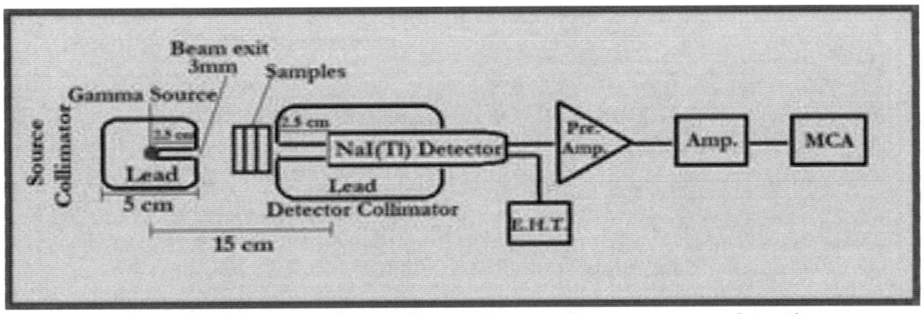

Figure1. Diagram of experimental setup for gamma ray detection.

2.4. Neutrons attenuation measurements

The BF_3 neutron detector was used to detect a collimated total slow neutron (main slow neutron emitted from the source and that slowed down by samples), slow neutron, and neutrons with energy higher than 10 keV beam released from ^{241}Am-Be neutrons source with activity 3.7 GBq. The neutrons transmitted flow were calculated according to equation (4) [28] to deduce the amounts of macroscopic neutron cross-section. The collimated beam was decelerating with polyethylene block of 7 cm thickness back the sample for slow neutrons calculations, and the neutrons that have energy below 10 keV were cut by a mass of Born carbide B_4C. Scheme of the experiment is shown in Figure 2.

$$I_n = I_0 e^{-\Sigma x} \qquad (4)$$

where Σ is the macroscopic cross section of neutrons, I_n and I_0 are the neutron fluxes after and before transmitted through the sample barriers respectively, and x is the sample thickness.
The maximum experimental errors in attenuation coefficients and half value layers for both gamma-ray and neutron attenuation measurements were evaluated using the following error formulas [29]:

$$\Delta\sigma = \frac{1}{x\rho}\sqrt{\left(\frac{\Delta I_0}{I_0}\right)^2 + \left(\frac{\Delta I_\gamma}{I_\gamma}\right)^2 + \left(Ln\frac{I_0}{I_\gamma}\right)^2\left(\frac{\Delta x\rho}{x\rho}\right)^2} \qquad (5)$$

$$(\Delta HVL)_\gamma = (HVL)_\gamma\sqrt{\left(\frac{\Delta\mu}{\mu}\right)^2(\ln 2)^2} \qquad (6)$$

$$\Delta\Sigma = \frac{1}{x}\sqrt{(\frac{\Delta I_0}{I_0})^2 + (\frac{\Delta I_n}{I_n})^2 + (Ln\frac{I_0}{I_n})^2(\frac{\Delta x}{x})^2} \quad (7)$$

$$(\Delta MFP)_n = (MFP)_n\sqrt{(\frac{\Delta\Sigma}{\Sigma})^2} \quad (8)$$

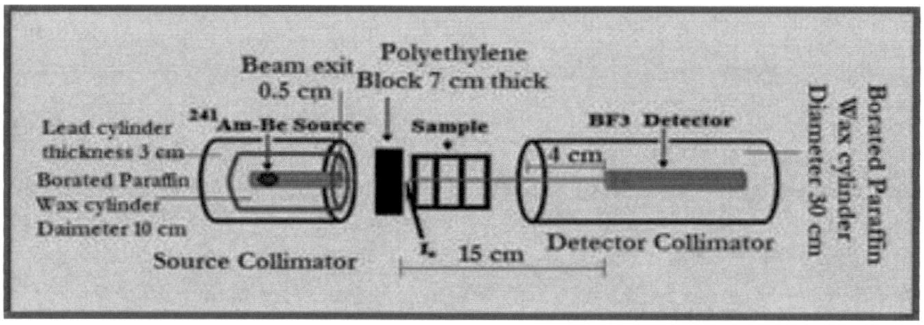

Figure2. Diagram of Neutron spectrometer.

3. Results and discussion

3.1. Mechanical properties

The mechanical properties (Vickers hardness, yield strength, ultimate tensile strength, elongation, and impact energy) of the investigated stainless steels were evaluated to calculate the effect of aluminum and tungsten contents on stainless steels. Table 3 and Figure 3 show the different mechanical properties for the four investigated stainless steel alloys.

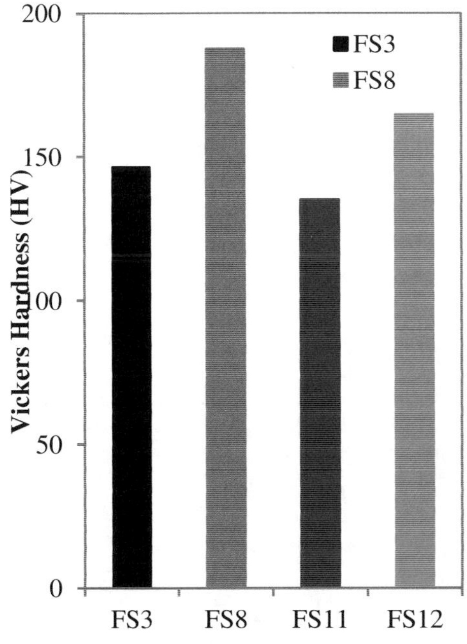
200
150
100
50
0
Vickers Hardness (HV)
FS3
FS8
FS3
FS8
FS11
FS12
Sample Code
(a)

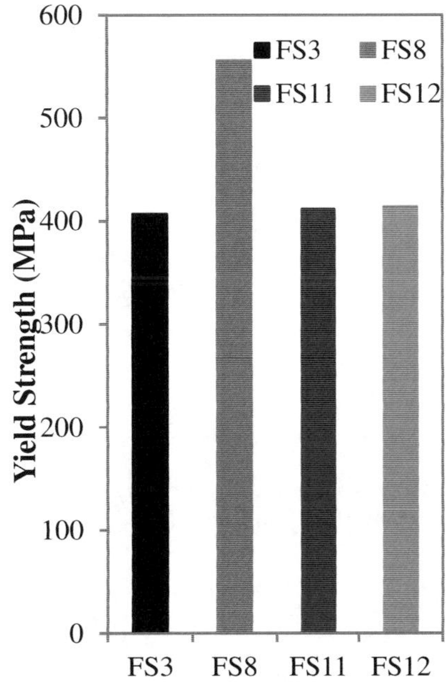
600
500
400
300
200
100
0
Yield Strength (MPa)
FS3
FS8
FS11
FS12
FS3
FS8
FS11
FS12
Sample Code
(b)

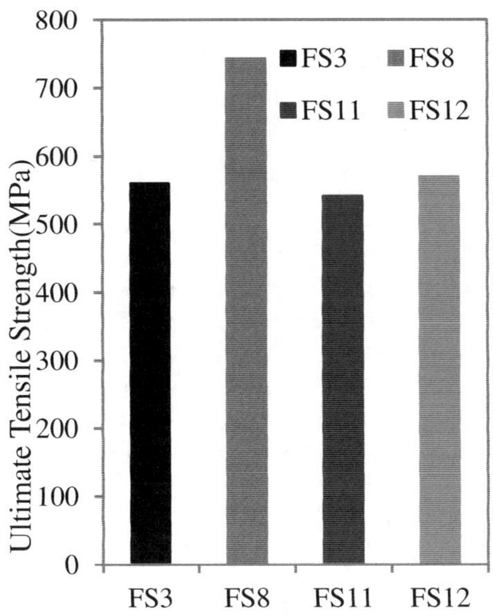
800
700
600
500
400
300
200
100
0
Ultimate Tensile Strength(MPa)
FS3
FS8
FS11
FS12
FS3
FS8
FS11
FS12
Sample Code
(c)

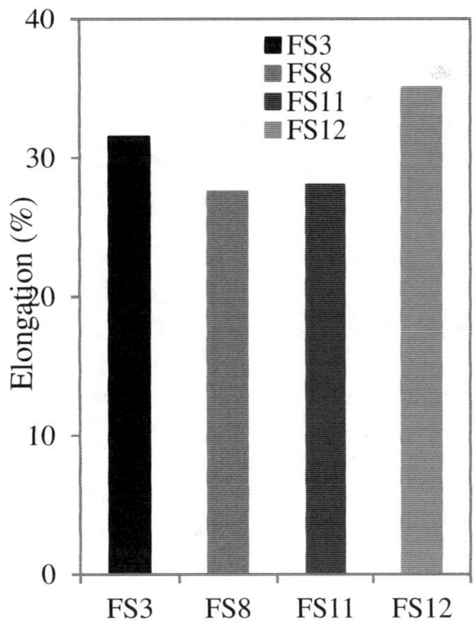
40
30
20
10
0
Elongation (%)
FS3
FS8
FS11
FS12
FS3
FS8
FS11
FS12
Sample Code
(d)

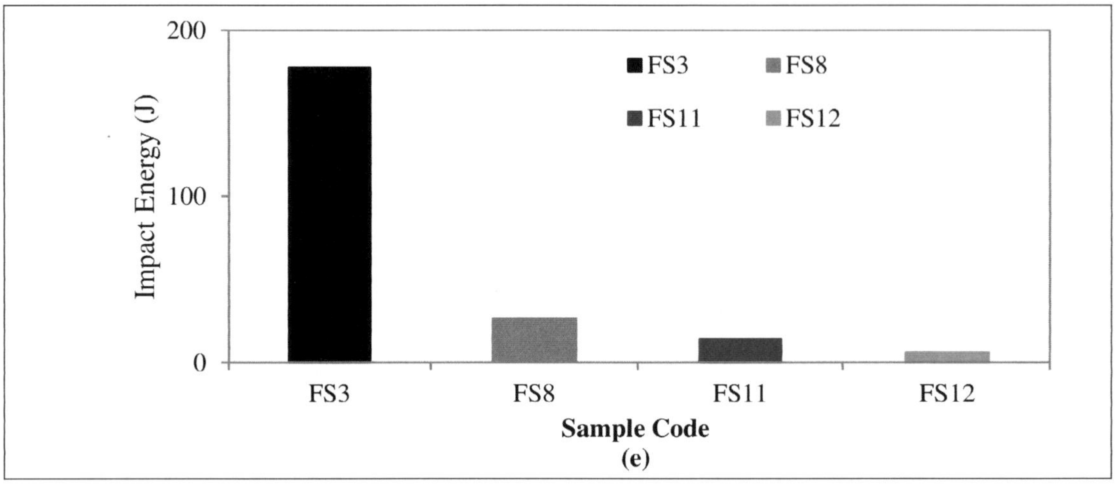

Figure 3. Mechanical properties of the investigated stainless steel alloys; (a) Vickers hardness, (b) yield strength, (c) ultimate tensile strength, (d)elongation, (e) impact energy.

Table 2. Mechanical properties of the investigated stainless steel alloys; hardness, yield strength, ultimate tensile strength, elongation, and impact energy.

Steel Code	Mechanical Properties				
	Vickers hardness (MPa)	Yield strength (MPa)	Ultimate tensile strength (MPa)	Elongation (%)	impact energy (J)
FS3	146.500	406.500	559.500	31.500	177.500
FS8	187.533	555.500	743.000	27.500	26.500
FS11	135.133	411.000	541.000	28.000	14.000
FS12	164.800	414.000	570.000	35.000	6.000

From the results shown above in Table 2 and represented in Figure 3, it was observed that yield and ultimate strength commonly depicts the structures of the material. Modified ferretic stainless steel alloys containing Al (FS8) exhibit higher hardness, yield strength and ultimate tensile strength than the standard FS3 and the other studied stainless steel alloys. This strengthening is accompanied with decreasing of elongation combined with reduction in impact energy.

As increasing the fine precipitation in the matrix, yield and ultimate strength is strongly enhanced. This refers to the interaction of fine precipitates with dislocations through pile up mechanism [31]. M_7C_3 which can be formed through solution treatment process of FS8 steel, due to its high carbon content. Hence, these precipitates a great impact on promoting the strength of steel. At the meantime, the coarsening of laves based tungsten Fe_2W has not the capability to interact with the dislocation, regarding to its coarse character. At the meantime, the laves in FS11, and FS12 has not a deteriorative effect on the ductility of ferrite matrix, in comparing with the reference steel FS3. As aforementioned, the toughness of the material is mainly depending on the fraction of secondary phases that are precipitated either throughout or along the boundaries. Secondary phase acts as stress concentration location that promote the propagation of the crack along the matrix [32]. Then, it is believed that volume fraction and size of secondary phase are the main parameter in deterioration of toughness. In fact, this concept was strongly confirmed on studying the toughness of investigated steel, as given in Table 2.

3.2. Gamma-ray measurements

A Co-60, Cs-137, and Eu-152 Gamma-ray origins were applied to calculate σ of gamma-radiation with energies (344.2, 661.6, 778.90, 964, 1112.4, 1173.2, 1332.5 and 1407.2 keV). The concentration of γ-radiations transmitted through alloy barriers were given as a function of surface concentrations (x_p, g/cm^2) at various percentage of Al and W to stainless steel alloy. The average values of (σ, cm^2/g) were derive as a slope of those exponential curves. Figure 4 shown measured and calculated results of energy dependence of the calculated and measured values of (σ_{calc}.& σ_{exp}.) for different stainless-steel alloys. Firstly, these figures display that a perfect agreement between experimental data of mass attenuation coefficients and that calculated by the "WinXcom" computer program (version 3.1). The action of all these curves can be interpreted by the prominent interaction between the studied alloy barriers and gamma rays which are considered a Compton scattering process. Where the total attenuation coefficient decreases quite slowly with rising photon energy E, where σ is inversely proportional to E. The value of (σ, cm^2/g) almost remain unaffected by adding Al or W to stainless steel alloy. This may be assigned to the z-dependence of the interaction of gamma rays with basically aluminum and tungsten. The comparison of half value layers of various types of investigated stainless steel alloys with and without Al,W for different used gamma ray energies are given in Table 3.

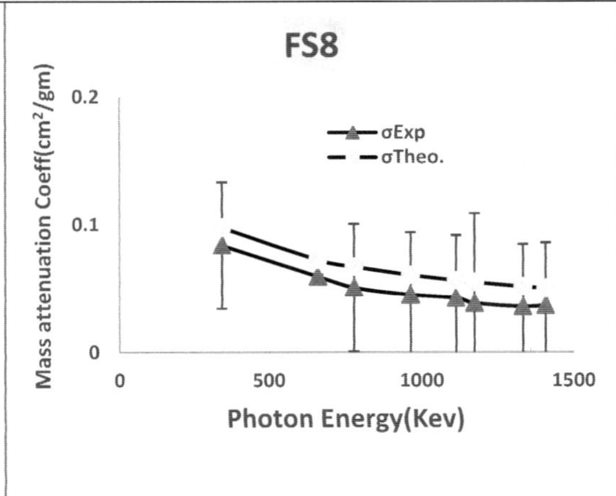

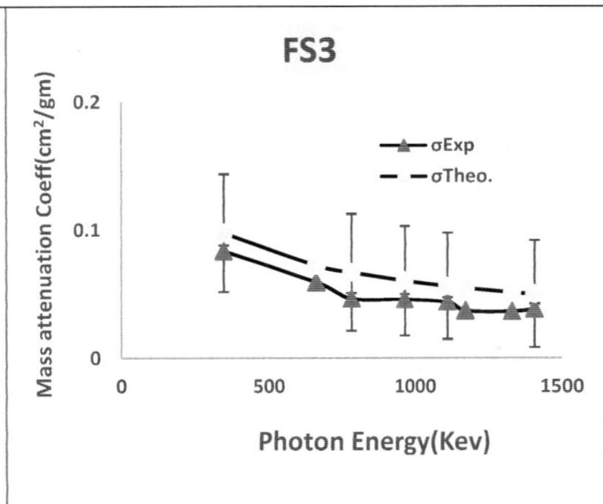

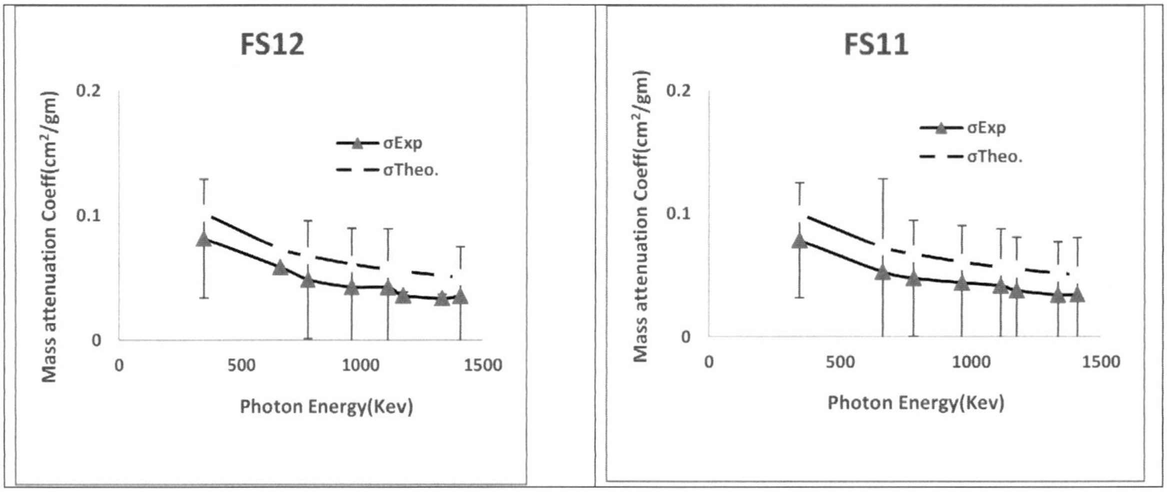

Figure 4. Experimental and calculated mass attenuation coefficients of the studied alloys.

Table 3. Half value layer (HVL) of the developed samples as a function of different gamma ray energies.

Energy(KeV)	HVL(cm)			
	FS3	FS8	FS11	FS12
344.14	1.065	1.066	1.118	1.063
661.64	1.500	1.504	1.662	1.484
778.90	1.920	1.763	1.839	1.786
964.00	1.942	1.986	1.998	2.045
1112.10	2.045	2.094	2.113	2.057
1173.23	2.415	2.326	2.334	2.424
1332.51	2.458	2.502	2.577	2.596
1407.9	2.333	2.432	2.558	2.458

3.3. Neutron measurements

^{241}Ame-Be neutron source wich activity 100 mCi was applied to determine the amounts of the macroscopic cross-section for total slow neutron, Σ_T, (main slow and go slow in the developed alloys), slow (Σ_S), and neutrons with energy higher than 10 keV ($\Sigma_{>10keV}$) in developed stainless steel alloys with percentages of Al and W. The values of (Σ, cm^{-1}) were measure as a slope of exponential curves, that the densities of neutrons with various energies transmitted during stainless steel alloy barriers given as a function of thicknesses at various percentages of Al and Won stainless steel alloy. Figure 5, shows the variations in values of Σ with distinct stainless-steel alloys. The amounts of macroscopic cross-sections of slow neutrons in whole kinds of stainless-steel alloys under investigation were the largest values of the others neutron energies. The behaviour of slow neutrons may be referred to elastic scattering of slow neutron with light elements in all forms of stainless steel alloys.

Also, it was noticed that the amounts of macroscopic cross-sections of total slow neutrons slightly increasing than that of neutron with energy higher than 10 keV in all studied alloys. This behavior may be attributed to the competition between absorption operation and slowdown process which takes place in the case of total slow neutron, while in the case of neutron with energy greater than 10 keV the only slow down process that appear. (note that, about 23% of neutron energies carried out from ^{241}Am-Be neutron source is considered slow neutron). Furthermore, Figure 6 shows the different values of mean free path, MFP, for investigated samples by three various energies of neutrons.

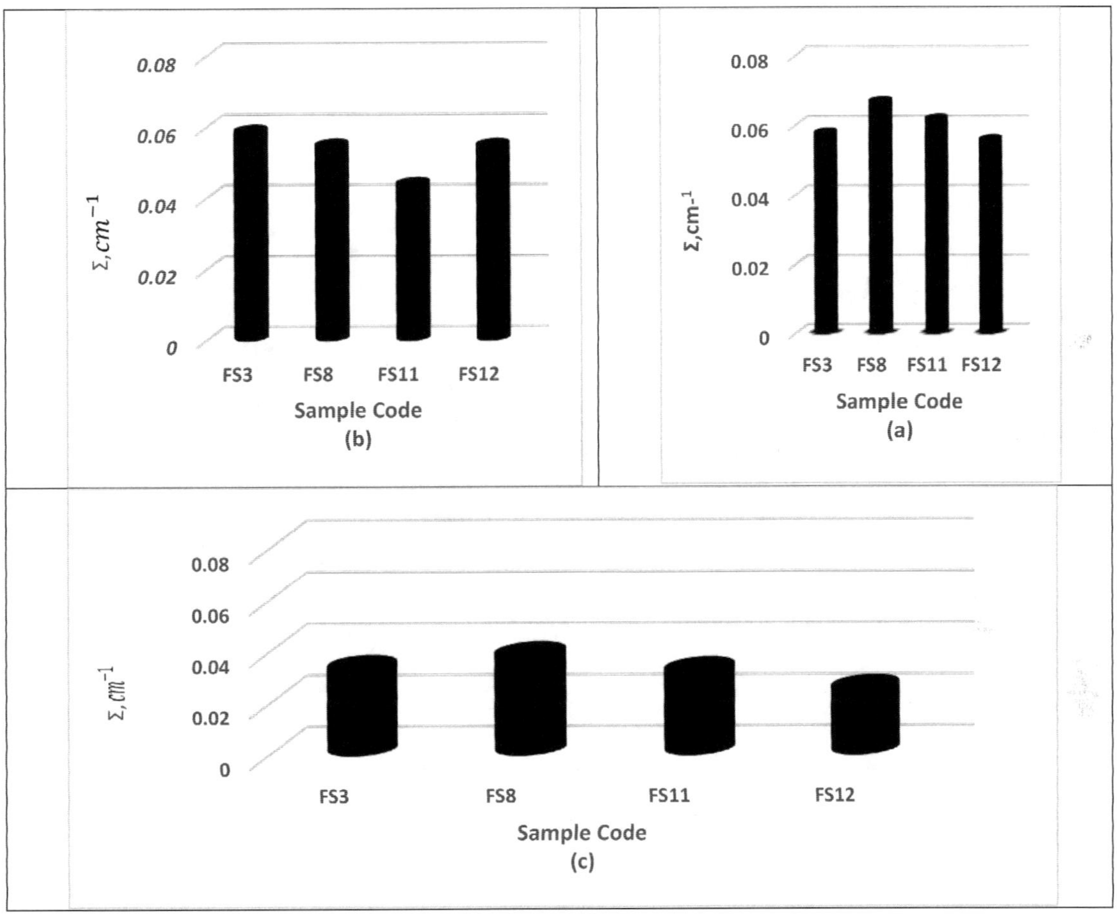

Figure 5. Macroscopic cross-sections of studied stainless steel alloys at (a) Total slow neutrons, (b) Slow neutrons, (c) Neutrons with energy > 10 KeV.

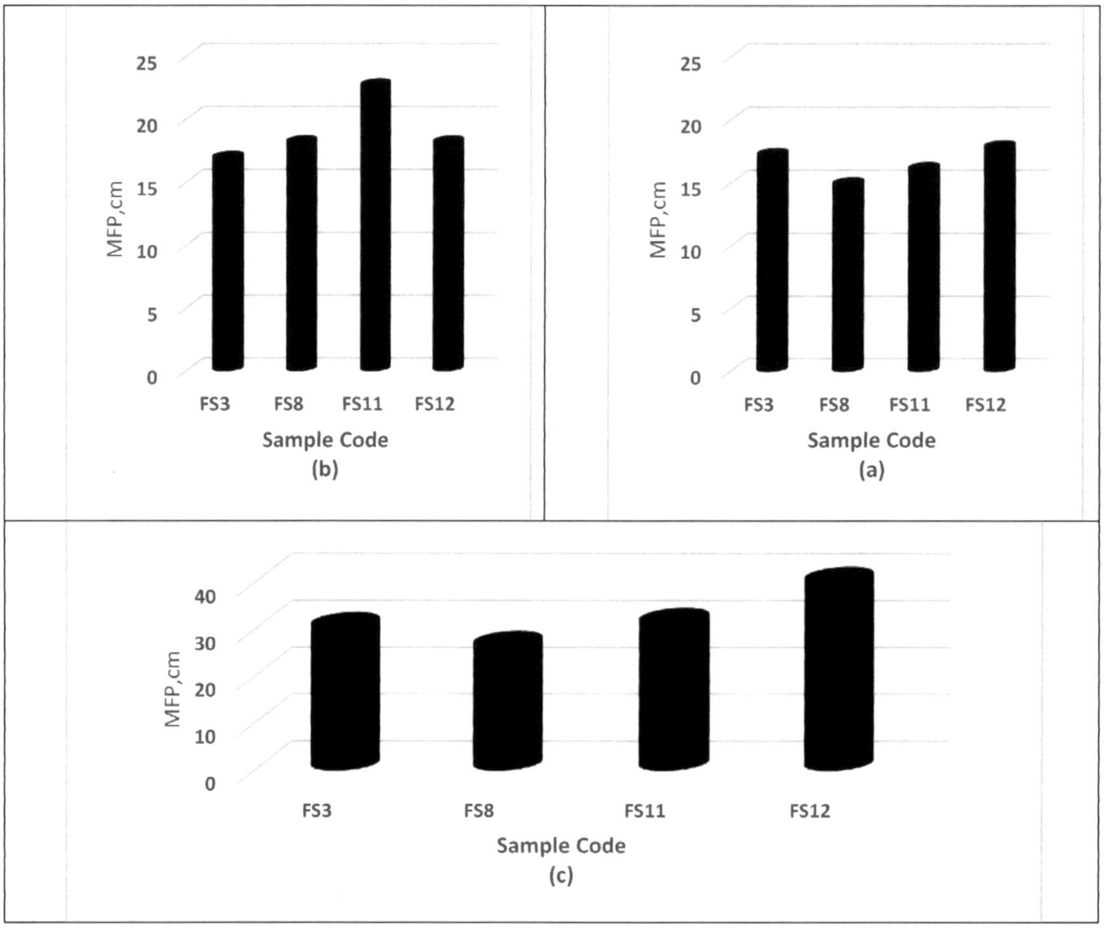

Figure 6. MFP of investigated samples for (a) total slow neutrons, (b) Slow neutrons, (c) Neutrons with energy > 10 KeV.

4. Conclusion

In the present study, four grades of stainless steel were developed and examined to use as candidate materials for nuclear reactor system.

The effect of Al and W can be assigned to the solution hardening and precipitation strengthening effects of these elements as these elements are strong carbide formers which accelerates the forming of intermetallic compounds and lowers the ductility.

The measured mass attenuation coefficients of gamma radiations are almost the same for all the developed alloys. A comparison of the mass attenuation coefficient with the corresponding theoretical one (based on XCOM) has been done and a good agreement is achieved.

The calculated macroscopic neutron cross-section of total slow, slow neutrons and neutrons with energy > 10 keV for the developed aluminum-tungsten stainless steel has a small variation compared with the standard sample FS3. This change could be assigned to the increase of the aluminum or tungsten contents in the studied alloys.

Achieved results reveal the superiority of the investigated alloys as a blanket structural substance for fusion reactor.

References

[1] International Atomic Energy Agen1995 Energy from inertial fusion **944**.
[2] Holdren, J. P., Berwald, D. H., Budnitz, R. J., Crocker, J. G., Delene, J. G., Endicott, R. D., ... and Schultz, K. R. 1988 Fusion Technology **13**(1) 7-56.
[3] Hadaller, O. J., Momenthy and A. M. 1993 Transportation and Global Climate Change Washington .
[4] Bolt, H., Barabash, V., Krauss, W., Linke, J., Neu, R., Suzuki, S., ... and Team, A. U., 2004 Journal of nuclear materials **329** 66-73.
[5] Şahin, S., Übeyli and M. 2008 Journal of fusion energy **27**(4) 271-277.
[6] Übeyli, M., Yalçın and Ş. 2006 Journal of Fusion Energy **25**(3-4) 197-205.
[7] Ihli, T., Basu, T. K., Giancarli, L. M., Konishi, S., Malang, S., Najmabadi, F., and Wu, Y., 2008 Fusion Engineering and Design **83**(7-9) 912-919.
[8] Brnic, J., Turkalj, G., Canadija, M., Lanc, D., and Krscanski S. 2011 Mechanics of time dependent materials **15**(4) 341-352.
[9] Hayward, T. M., Svishchev, I. M., Makhija, and R. C. 2003 supercritical fluids **27**(3) 275-281.
[10] Singh, N., Singh, K. J., Singh, K., & Singh, and H., 2004, Interactions with Materials and Atoms **225**(3) 305-309.
[11] Singh, V. P., & Badiger, and N. M. 2014 Annals of Nuclear Engineering **64** 301-310.
[12] Eissa, M. M., El-Kameesy, S. U., El-Fiki, S. A., Ghali, S. N., El Shazly, R. M., and Saeed A. 2016 Fusion Engineering and Design **112** 130-135.
[13] Byun, T. S., Hashimoto, N.,and Farrell K. 2004 Acta Materialia **52**(13) 3889-3899.
[14] Min, K. D., Hong, S., Kim, D. W., Lee, B. S., Kim, and S. J. 2017 Nuclear Engineering and Technology **49**(4) 752-759.
[15] Fuzeau, J., Vasudevan, M., and Maduraimuthu V. 2016 Transactions of the Indian Institute of Metals **69**(8) 1493-1499.
[16] Kim, J. K., Kim, Y. H., Lee, J. S., Kim, and K. Y. 2010 Corrosion Science **52**(5) 1847-1852.
[17] Liu, H., Wei, L., Ma, M., Zheng, J., Chen, L., Misra, and R. D. K. 2020 Materials research and Technology **9**(2) 2127-2135.
[18] Brady, M. P., Yamamoto, Y., Muralidharan, G., Rogers, H., Pint, and B. A. 2013 Oak Ridge TN.
[19] Meetham, G. W., Van de Voorde, and M. H. 2000 Springer Science & Business Media.
[20] Salama, E., Eissa, M. M., Tageldin, and A. S. 2019 Nuclear Engineering and Technology **51**(3) 784-791.
[21] Zhang, X., Fan, L., Xu, Y., Li, J., Xiao, X., and Jiang, L. 2015 Materials & Design **65** 682-689.
[22] Jang, M. H., Moon, J., Kang, J. Y., Ha, H. Y., Choi, B. G., Lee, T. H., Lee, and C. 2015 Materials Science and Engineering: A **647** 163-169.
[23] Klueh, R. L., Maziasz, and P. J. 1989 Metallurgical Transactions A **20**(3) 373-382.
[24] Zhao, J., Lee, T., Lee, J. H., Jiang, Z., Lee, and C. S. 2013 Materials and Metallurgical Transactions **44**(8) 3511-3523.
[25] Prifiharni, S., Anwar, M. S., Nikitasari, A., Mabruri, and E. 2018 In AIP Conference Proceedings **1964**(1) 20041 AIP Publishing LLC.
[26] Lamarsh, J. R., Baratta, and A. J. 2001 **3** 783 Upper Saddle River, NJ: Prentice hall.
[27] El Shazly, R. M., and M. M. Sadawy. 2017 International Journal of Scientific Engineering and Research (IJSER) **5**: 243-250.
[28] Saeed, A., Elbashar, Y. H., Abou El-azm, A. M., El-Okr, M. M., Comsan, M. N. H., Osman, A. M., El-Sersy, and A. R. 2014 Rad. Phy. and Chem. **102** 167-170.
[29] Taylor, and J. 1997 the study of uncertainties in physical measurements .

Pulsed versus continuous-wave Laser deoxygenation of graphene oxide suspensions

T. A. El-Brolossy*, T. Abdallah, H. Talaat

Physics Department, Faculty of Science, Ain shams University, Cairo, Egypt.

* E-mail: elbrolosyta@gmail.com

Abstract. We report on a comparative study of laser deoxygenation of graphene oxide (GO) suspensions in pure water (pH~7) using pulsed (nanosecond with wavelength 532 nm and fluence ~ 0.15 J/cm^2) and continuous-wave (CW) (of wavelength 448 nm and power density ~ 0.25 MW/cm^2) lasers at room temperature. Such comparison was not performed previously for GO suspensions. Concentrated GO colloidal suspensions were prepared by a modified Hummer's method. The presence of oxide groups in GO was proved using X-ray diffraction and Fourier transform infrared (FTIR). The laser deoxygenation processes were verified by means of UV–visible, FTIR and micro-Raman spectroscopies. The impact of changing the laser irradiation time for both lasers has been investigated using the characteristic Raman bands (D and G) of GO as markers to assess the degree of reduction as well as the structural defects of the resulting laser RGO.

Keywords: Graphene oxide suspensions, laser reduction, Raman spectroscopy, reduced graphene oxide

1. Introduction

Graphene oxide (GO) is one of the most significant graphene byproducts utilized for mass production of graphene [1-4]. Its structure is like a graphene and characterized by a random allocation of oxygen functional groups attached at sp3 hybridized carbon atoms (C-sp^3) in what is called oxidized area. Merged with this region are non-oxidized domain constituted by hydrophobic π-conjugated C-sp^2. The former region contains epoxy and hydroxyl groups at the basal planes, while edges are linked to carboxyl and carbonyl groups [5]. This structure of GO makes it of poor electrical conductivity that restricts its applications in microelectronics. Deoxygenation of GO affords another pathway to graphene-like sheets [6]. GO can be reduced chemically with reducing agents [7] or through annealing at relatively high temperatures [8], as well as using photo-thermal reduction, or elective reduction by straightforward laser writing [9, 10]. Selection of the reduction method of GO is decisive where it determines the properties of the resultant reduced graphene oxide (RGO) material [11].

The laser deoxygenation of GO allows creation of conductive domains with improved optical properties, which can be appropriate in photo-electronics devices, such as in organic photovoltaic devices, and biological sensors etc. [12-15] . As well, laser reduction of GO is appealing owing the ability to control the extent of reduction process depending on the experimental parameters as GO concentration and irradiation conditions. In addition, this method has an advantage like other photo-irradiation processes, that it does not depend on utilizing chemicals or high temperatures. The production of a homogeneous colloidal suspension of electrically and thermally conductive graphene sheets in solvents that allows their broad use for both essential examination and in applications is a basic interest [16-19]. However, studies concerning the use of lasers in the reduction processes of GO suspensions are mainly through pulsed laser of high power [20, 21]. Continuous-wave (CW) laser never used before in the reduction process of GO suspensions [22].

In this work, we provide visible laser reduction of GO suspensions by both pulsed and CW laser. The method offers a suspension production of individual RGO sheets in water under ambient conditions. The structure of GO was investigated and characterized using X-ray diffraction (XRD), UV-Vis, Fourier transform infrared (FTIR) spectroscopy as well as micro-Raman spectroscopy. Micro-Raman spectroscopy is a powerful technique for studying crystallographic defects and local stress distribution [23, 24]. Also this spectroscopy is omnipresent technique for examining sp^2 domain size of GO, lattice order, functionalization, and sheet-stacking arrangement [25]. Therefore, we have employed Raman spectroscopy to assess and compare the impact of irradiation time on the reduction process by both pulsed and CW lasers

2. Experimental

2.1 GO preparation

Modified Hummer's method was used to prepare GO [26]. In a typical method, 46 mL of concentrated H_2SO_4 was mixed up with 1 g graphite powder with stirring in an ice bath. To fix the mixture temperature lower than 20 ^oC, the oxidizing agent KMnO4 (3.0 g) was added bit by bit under strong stirring. Consecutively, the ice bath was removed, and the reaction system was vigorously stirred for 0.5 h at about 35 ^oC. Next, 100 mL water was added up gently, and the solution was stirred for 15 min at 95 ^oC. Then, the solution was poured into 250 mL water followed by a dropwise addition of 5 mL H_2O_2 (30%), changing the color of the solution from dark brown to yellow brown. The suspension was held uninterrupted overnight where GO precipitated. The washing stage starts with removing excess metal ions by adding 1:9 HCl aqueous solution (250 mL) to the precipitated GO and centrifuged at 7000 rpm. Recurrent washing using distilled water was performed until the pH was ~7. The resulting GO precipitates were dried at room temperature for 24 h.

2.2 Laser reduced GO

A yellow homogeneous dispersion of GO was obtained by sonicating 1 mg of dried GO powder in 10 mL of deionized water. Two experiments for GO suspension irradiation were conducted using two different lasers. The first is a second harmonic (λ= 532 nm) pulsed Nd/YAG laser of pulse width 9 ns and repetition rate = 20 Hz at low pulse fluence ~ 0.15 J/cm^2 whereas the second is a CW laser of λ= 448 nm at power density 0.25 MW/cm^2. The irradiation time (τ) varies according to the laser used, starting at 5 up to 60 minutes for 532 nm lasers, while for 448 nm laser the irradiation time begins at 60 minutes up to 300 minutes.

2.3 *Characterization*

XRD pattern of GO was collected at room temperature by a Philips diffractometer (X'pert MPD) using Cu Kα (λ= 1.54 A^o) radiation source. Absorption spectra were picked up employing UV-Vis spectrophotometer. Oxidation and deoxygenation processes were assessed by FT-IR spectroscopy using Bruker alfa II system in the range 500 to 4000 cm^{-1} in the transmission mode by the aid of KBr (IR grade). The structures of the GO and RGO were studied by micro-Raman spectroscopy (LabRam HR Evolution-Horiba spectrometer) using 532 nm laser line as an excitation source.

3. Results and Discussion

The XRD pattern (Lorentzian fit) of the dried GO powder is shown in Fig. 1. The diffractogram shows a sharp peak centered at 2θ = 11.5°, which is a characteristic peak of GO corresponding to the (001) inter-planar (d) spacing of 7.69 Å calculated by Bragg's equation. This value of interlayer distance is larger than twice of the original graphite (3.34 Å) [27]. This increase of inter-planar distance is attributable to the entrapping of oxygen functional groups between the graphene oxide sheets confirming the oxidation process [28]. The broad peak centered around 2θ ~ 26° can be attributed to the residual non-oxidized graphite, while the weak peak centered at 2θ ~ 42° is related to the in-plan crystallinity.

Figure 2 (a – c) shows the UV–vis spectra of unirradiated, CW laser irradiated (τ = 300 min) and pulsed laser irradiated (τ = 60 min) GO suspensions, respectively. The position of the absorption peak of the spectra is deemed as an indication of the deoxygenation degree. The $\pi - \pi^*$ transition in the aromatic bands is clearly observed in the GO suspension spectrum through the strong absorption peak at ~231 nm [29]. Also, the existence of the carbonyl groups is verified by the obvious shoulder at about 300 nm corresponding to $n \rightarrow \pi^*$ transitions. The laser irradiation provokes a reclamation of the electronic conjugation inside the RGO sheets

[30]. This is in harmony with the change in color of the suspension from yellow for the GO to black for the CW and pulsed laser-RGO obtained after 300- and 60-min irradiation respectively, as displayed in the inset of Figure 2. Also, the red shift of the strong absorption peak from 231 to ~261 nm is caused by further $\pi - \pi^*$ transitions in the aromatic bands of the restored sheets which has become in need to less energy for the electronic transition. Furthermore, the disappearance of shoulder peak (~300 nm) in the spectra of the laser irradiated suspensions indicates the elimination of the carbocyclic and carbonyl groups from the surface of the laser irradiated GO sheets

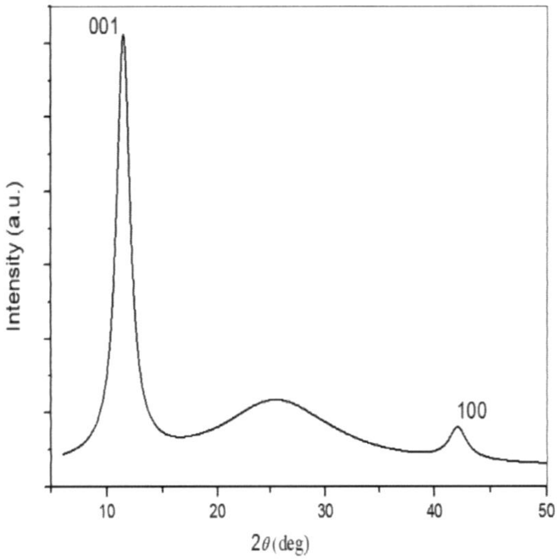

Fig. 1 XRD pattern of the prepared GO sheets

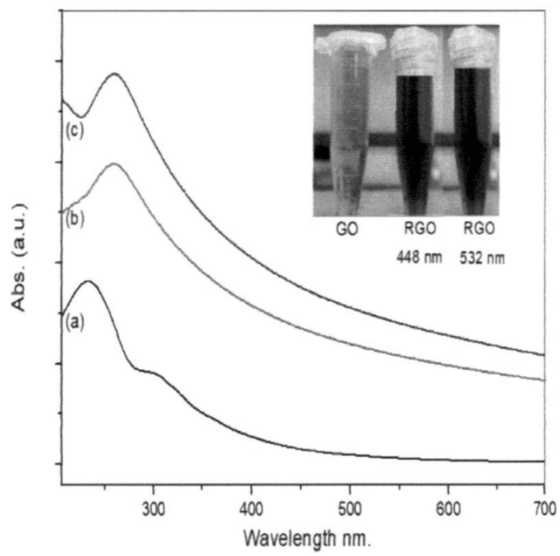

Fig. 2 UV–vis spectra of (a) unirradiated, (b) CW laser irradiated (τ = 300 min) and (c) pulsed laser irradiated (τ = 60 min) GO suspensions

Figure 3 shows FT-IR spectrum of GO (a) together with those spectra of RGO under the influence of laser irradiation with CW laser for 300 min (b) and pulsed laser for 60 min (c). The GO spectrum reveals a clear existence of oxygen containing functional groups as an evidence of the successful oxidation process. The characteristic bands include a broad peak at ~ 3420 cm^{-1} corresponding to the stretching of OH group [31], a strong band at 1725 cm^{-1} related to the C=O stretching vibrations of the carboxyl groups, and the C—O stretching of epoxide group at 1178 and1079 cm^{-1} [32]. The band which reflects the C=C stretching in graphene structure appears at 1630 cm^{-1}. The intensity of this band relative to the intensities of the characteristic bands corresponding to oxygen functional groups can be used to assess the deoxygenation process [31]. Most of these bands have entirely disappeared from the spectra of suspensions treated with both lasers at the specified irradiation time used for each as shown in Fig. 3 (b & c). Such results confirm the significant deoxygenation of GO following pulsed as well as CW laser irradiation under the previously mentioned conditions

The Raman spectrum of GO in the range 1100 – 1800 cm^{-1} is displayed in Fig. 4a. This spectrum shows the characteristic first order peaks (G and D bands) of the exfoliated GO. The one-phonon process of E_{2g} optical phonon known as G - band is attributed to the in-plane vibration of the C atom sp^2 network [33]. While the D-band is a breathing mode of A_{1g} symmetry induced by structural sp^3 defects [34]. The G-band appears at 1588 cm^{-1} which is broadened and blue-shifted from that of the graphite G-band shown in the inset of Fig. 4, The D-band that be seen at ~1350 cm^{-1} has a considerably higher intensity and broadening relative to that of graphite. The considerable broadening of the apparent G band could be referred to the existence of an overlapped defect activated band D'. This band has been ascribed to phonon confinement effect provoked by defects. Besides the D to G (actual band) intensity ratio (I_D/I_G) which is ~ 0.14 for graphite increases to be ~ 1.43 for GO. These pronounced variations are related to significant defects and structural disorder caused by various oxygen functional groups present in the basal plane and terminating the edges.

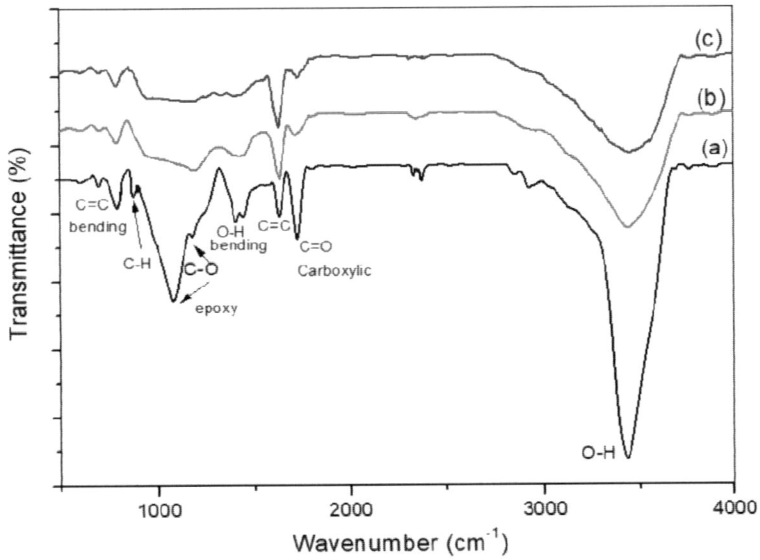

Fig. 3 FTIR spectra of (a) unirradiated, (b) CW laser irradiated (τ = 300 min) and (c) pulsed laser irradiated (τ = 60 min) GO

The Raman spectrum of the 448 nm CW laser irradiated GO (τ = 300 min) is presented in Fig. 4b. The spectrum shows a stronger and little resolved G – band (~1580 cm^{-1}) relative to that of GO without noticeable narrowing of the apparent band. This slight resolution allows a weak appearance of the $D^{/}$ band as a broad shoulder that indicate a small reduction of the width of the G band. The D-band also appears around 1350 cm^{-1} with reasonable intensity: however, I_D/I_G decreases to be ~ 0.96. Figure 4c shows the Raman spectrum of GO irradiated with 532 nm pulsed laser (τ = 60 min). There is a clear narrowing of the G band with further red shift (~1575 cm^{-1}) allows the observation of $D^{/}$ as a small shoulder around 1610 cm^{-1}. Furthermore, a pronounced reduction of D band width and intensity is observed in the Raman spectrum shown in Fig. 4c, where the ratio I_D/I_G decreases to be ~ 0.4. Generally, the narrowing of G band and D bands point to the existence of fewer defects in the structure because of lower oxygen content [35]. The intensity ratio I_D/I_G is used as a quality assessment of the of the graphitic structures, it reflects how extent is sp^2 domain. The variation of this ratio as a function of irradiation time τ for pulsed and CW laser irradiated GO are shown in Figure 5 (a & b) respectively. The continuing decrease of this ratio with increasing τ implies expansion of sp^2 domain which indicates more conversion of GO into graphene [36].

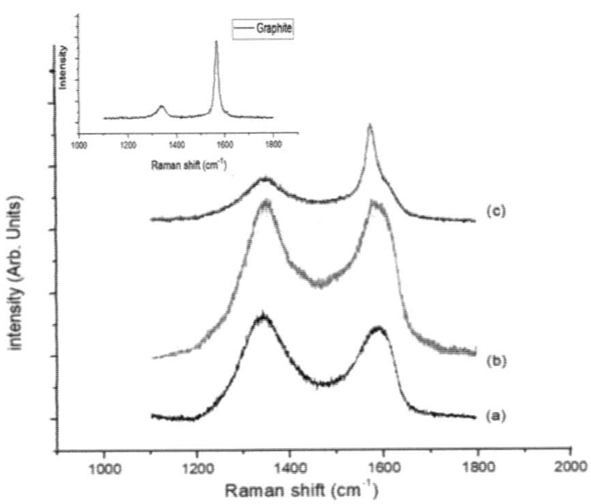

IOP Conf. Series: Materials Science and Engineering **956** (2020) 012009 doi:10.1088/1757-899X/956/1/012009

Fig. 4 Raman spectra of (a) unirradiated, (b) CW laser irradiated (τ = 300 min) and (c) pulsed laser irradiated (τ = 60 min) GO (the inset is the Raman spectrum of graphite)

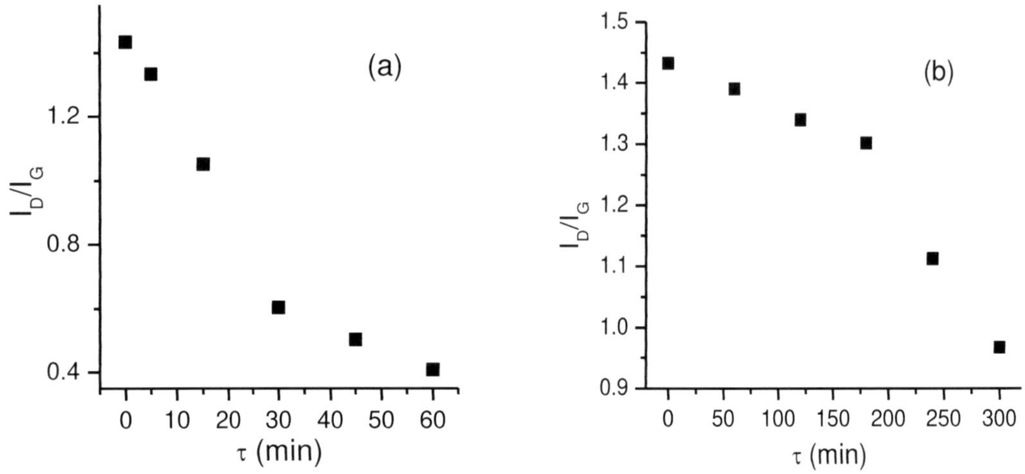

Figure 5 The variation of I_D/I_G ratio as a function of irradiation time for pulsed (a) and CW (b) laser irradiated GO

4. Conclusion

In conclusion, CW laser (448 nm) of relatively low power density (0.25 MW/cm^2) as well as pulsed laser (532 nm) of least pulse fluence (0.15 J/cm^2) used in this study can deoxygenate GO suspensions. The results of UV-Vis as well as FTIR reveal that under these conditions the deoxygenation process of GO using CW laser needs irradiation period up to 300 min to produce comparable results to those obtained using pulsed laser of 60 min irradiation time. However, Raman spectroscopy offers other comparisons to distinguish between RGOs produced due to irradiation by both lasers through their structural properties. The larger drop in I_D/I_G in case of pulsed laser reduced GO may be an indication to more efficient conversion of GO into graphene. More Raman investigations regarding comparing crystallite size, defect density and type of defects of the RGO suspensions formed by these lasers are going to be presented in the forthcoming work.

References

[1] Becerril H A, Mao J, Liu Z, Stoltenberg R M, Bao Z and Chen Y 2008 Evaluation of solution-processed reduced graphene oxide films as transparent conductors *ACS nano* **2** 463-70

[2] Huang X, Zhou X, Wu S, Wei Y, Qi X, Zhang J, Boey F and Zhang H 2010 Reduced graphene oxide-templated photochemical synthesis and in situ assembly of Au nanodots to orderly patterned Au nanodot chains *Small* **6** 513-6

[3] Huang X, Li S, Wu S, Huang Y, Boey F, Gan C L and Zhang H 2012 Graphene oxide-templated synthesis of ultrathin or tadpole-shaped Au nanowires with alternating hcp and fcc domains *Advanced Materials* **24** 979-83

[4] Wang Y-F, Xin F-F, Deng Y-R, Li D-J and Li X-F 2020 Nano-Zn2SnO4/Reduced Graphene Oxide Composites for enhanced photocatalytic performance *Materials Chemistry and Physics* **254** 123505

[5] Stankovich S, Dikin D A, Piner R D, Kohlhaas K A, Kleinhammes A, Jia Y, Wu Y, Nguyen S T and Ruoff R S 2007 Synthesis of graphene-based nanosheets via chemical reduction of exfoliated graphite oxide *carbon* **45** 1558-65

[6] Eda G, Lin Y-Y, Miller S, Chen C-W, Su W-F and Chhowalla M 2008 Transparent and conducting electrodes for organic electronics from reduced graphene oxide *Applied Physics Letters* **92** 209

[7] Moon I K, Lee J and Lee H 2011 Highly qualified reduced graphene oxides: the best chemical reduction *Chemical Communications* **47** 9681-3

[8] Larciprete R, Fabris S, Sun T, Lacovig P, Baraldi A and Lizzit S 2011 Dual path mechanism in the thermal reduction of graphene oxide *Journal of the American Chemical Society* **133** 17315-21

[9] Spanò S F, Isgrò G, Russo P, Fragalà M E and Compagnini G 2014 Tunable properties of graphene oxide reduced by laser irradiation *Applied Physics A* **117** 19-23
[10] Bobrinetskiy I, Emelianov A, Smagulova S, Komarov I, Otero N and Romero P 2017 Laser direct 3D patterning and reduction of graphene oxide film on polymer substrate *Materials Letters* **187** 20-3
[11] Pei S and Cheng H-M 2012 The reduction of graphene oxide *Carbon* **50** 3210-28
[12] Sokolov D A, Shepperd K R and Orlando T M 2010 Formation of graphene features from direct laser-induced reduction of graphite oxide *The Journal of Physical Chemistry Letters* **1** 2633-6
[13] Trusovas R, Ratautas K, Račiukaitis G, Barkauskas J, Stankevičienė I, Niaura G and Mažeikienė R 2013 Reduction of graphite oxide to graphene with laser irradiation *Carbon* **52** 574-82
[14] Kymakis E, Petridis C, Anthopoulos T D and Stratakis E 2013 Laser-assisted reduction of graphene oxide for flexible, large-area optoelectronics *IEEE Journal of Selected Topics in Quantum Electronics* **20** 106-15
[15] Malinský P, Macková A, Cutroneo M, Siegel J, Bohačová M, Klímova K, Švorčík V and Sofer Z 2018 Laser modification of graphene oxide layers. In: *EPJ Web of Conferences*: EDP Sciences) p 04010
[16] Huang X, Qi X, Boey F and Zhang H 2012 Graphene-based composites *Chemical Society Reviews* **41** 666-86
[17] Yuan W and Shi G 2013 Graphene-based gas sensors *Journal of Materials Chemistry A* **1** 10078-91
[18] Randviir E P, Brownson D A C and Banks C E 2014 A decade of graphene research: production, applications and outlook *Materials Today* **17** 426-32
[19] El-Brolossy T and Almutawa A 2020 Thermal Diffusivity and Thermal Effusivity Enhancement of Graphene based Polyvinyl Chloride Nanocomposites *MS&E* **762** 012006
[20] Abdelsayed V, Moussa S, Hassan H M, Aluri H S, Collinson M M and El-Shall M S 2010 Photothermal deoxygenation of graphite oxide with laser excitation in solution and graphene-aided increase in water temperature *The Journal of Physical Chemistry Letters* **1** 2804-9
[21] Ghadim E E, Rashidi N, Kimiagar S, Akhavan O, Manouchehri F and Ghaderi E 2014 Pulsed laser irradiation for environment friendly reduction of graphene oxide suspensions *Applied surface science* **301** 183-8
[22] Wang F, Wang K, Zheng B, Dong X, Mei X, Lv J, Duan W and Wang W 2018 Laser-induced graphene: preparation, functionalization and applications *Materials technology* **33** 340-56
[23] Martín E, Jiménez J and Chafai M 1998 MicroRaman study of crystallographic defects in SiC crystals *Solid-State Electronics* **42** 2309-14
[24] Talaat H, Negm S, Schaffer H E, Kaltsas G and Nassiopoulou A G 2000 Micro-Raman analysis of polysilicon membranes deposited on porous silicon channels *Journal of Non-Crystalline Solids* **266-269** 1345-9
[25] Díez-Betriu X, Álvarez-García S, Botas C, Álvarez P, Sánchez-Marcos J, Prieto C, Menéndez R and de Andrés A 2013 Raman spectroscopy for the study of reduction mechanisms and optimization of conductivity in graphene oxide thin films *Journal of Materials Chemistry C* **1** 6905-12
[26] Chen J, Yao B, Li C and Shi G 2013 An improved Hummers method for eco-friendly synthesis of graphene oxide *Carbon* **64** 225-9
[27] Mishra S K, Tripathi S N, Choudhary V and Gupta B D 2014 SPR based fibre optic ammonia gas sensor utilizing nanocomposite film of PMMA/reduced graphene oxide prepared by in situ polymerization *Sensors and Actuators B: Chemical* **199** 190-200
[28] Saleem H and Habib A 2016 Study of band gap reduction of TiO2 thin films with variation in GO contents and use of TiO2/Graphene composite in hybrid solar cell *Journal of Alloys and Compounds* **679** 177-83
[29] Luo Z, Lu Y, Somers L A and Johnson A T C 2009 High Yield Preparation of Macroscopic Graphene Oxide Membranes *Journal of the American Chemical Society* **131** 898-9
[30] Li D, Müller M B, Gilje S, Kaner R B and Wallace G G 2008 Processable aqueous dispersions of graphene nanosheets *Nature Nanotechnology* **3** 101-5
[31] El Achaby M, Arrakhiz F, Vaudreuil S, Essassi E and Qaiss A 2012 Piezoelectric β-polymorph formation and properties enhancement in graphene oxide–PVDF nanocomposite films *Applied Surface Science* **258** 7668-77
[32] Thakur S and Karak N 2012 Green reduction of graphene oxide by aqueous phytoextracts *Carbon* **50** 5331-9
[33] Ferrari A C, Meyer J, Scardaci V, Casiraghi C, Lazzeri M, Mauri F, Piscanec S, Jiang D, Novoselov K and Roth S 2006 Raman spectrum of graphene and graphene layers *Physical review letters* **97** 187401
[34] Ferrari A C 2007 Raman spectroscopy of graphene and graphite: Disorder, electron–phonon coupling, doping and nonadiabatic effects *Solid state communications* **143** 47-57

IOP Conf. Series: Materials Science and Engineering **956** (2020) 012009 doi:10.1088/1757-899X/956/1/012009

[35] Muzyka R, Drewniak S, Pustelny T, Chrubasik M and Gryglewicz G 2018 Characterization of graphite oxide and reduced graphene oxide obtained from different graphite precursors and oxidized by different methods using Raman spectroscopy *Materials* **11** 1050
[36] Pimenta M, Dresselhaus G, Dresselhaus M S, Cancado L, Jorio A and Saito R 2007 Studying disorder in graphite-based systems by Raman spectroscopy *Physical chemistry chemical physics* **9** 1276-90

IOP Conf. Series: Materials Science and Engineering **956** (2020) 012010 doi:10.1088/1757-899X/956/1/012010

Tunable filter based on the 1D photonic crystal within ultraviolet radiations

A. H. Aly[1], S. Barakat[2] and A. F. Amin[2]

[1]**TH-PPM group, Physics Department, Faculty of Sciences, Beni-Suef University, Egypt**

[2]**Faculty of Technology and Education, Beni-Suef University, Beni-Suef, Egypt**

Abstract

In the present study, we develop a one-dimensional semiconductor photonic crystal to control ultraviolet radiation. We study the transmittance spectra of incident light on the proposed structure in the ultraviolet regime. A photonic bandgap appears with a width of 95.2 nm. by changing the thickness parameters, it can reach up to 125.6 nm with appearing of new band gaps. The incident angle effect on the band gap's width is apparent. The bandgap shifted to a lower wavelength region. So the proposed structure could be useful in preventing UV radiation from passing through it.

Introduction:

According to the American cancer society, exposure to UV rays in sunlight causes the most skin cancers and other health problems such as sunburn, premature aging of the skin, and eye problems [1]. So in the last few years, scientists have been working on controlling such radiation by using several structures. The photonic crystal structures are one of the essential structures used in optical filter applications due to its remarkable features. Photonic crystals are artificial structures with periodic modulated dielectric constants in one, two, and three dimensions [2-11]. One of the essential features of the photonic crystal structures is the presence of the photonic bandgap. The photonic bandgap is the region where the incident light is prohibited and cannot propagate through the structure [12-17]. This phenomenon exists due to the periodic variation of the refractive indices at the structure interface [18]. Semiconductor photonic crystals have attracted extensive attention due to their remarkable physical properties [19]. ZnSe is a significant semiconductor material with a bandgap of 2.70 eV [20]. The ZnSe thin-film exhibits many physical phenomena with novel optical properties; it is also applicable for buffer/window [21-24]. In this work, we designed a one-dimensional semiconductor photonic crystal structure consisting of two semiconductor material in two layers and studied its optical properties using the transfer matrix method.

Theoretical analysis:

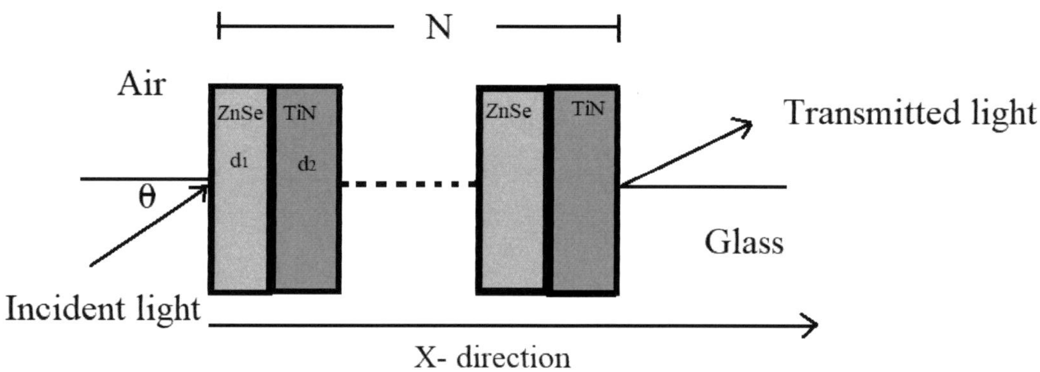

Figure (1): schematic diagram of our one-dimensional photonic crystal structure consisting of two semiconductor materials ZnSe, and TiN for N periods.

The geometry of one-dimension semiconductor photonic crystal is shown in figure.1. This structure consists of two layers repeated for N periods at N=15, the first layer is ZnSe material with thickness d_1, and the second layer is from TiN material with thickness d_2. The refractive index of ZnSe is n_1=2.44[25], and the refractive index of is TiN n_2=1.56[26]. The interaction between electromagnetic waves and the current structures can be obtained based on the analysis of the electric field interaction within the structure along the specified direction in terms of the dynamical and propagating matrices [17, 27]:-

$$F=\begin{pmatrix} f_{11} & f_{12} \\ f_{21} & f_{22} \end{pmatrix} = D_o{}^{-1}(D_1P_1D_1{}^{-1}D_2P_2D_2{}^{-1})^N D_s \quad (1)$$

$$D_m = \begin{bmatrix} 1 & 1 \\ n_m\cos\theta_m & -n_m\cos\theta_m \end{bmatrix} \quad (2)$$

At, $m = 0$, 1 and 2. Wherein, the propagation matrices (P) take the form:-

$$P_{m=}\begin{bmatrix} cos\varphi_l + isin\varphi_l & 0 \\ 0 & cos\varphi_l - isin\varphi_l \end{bmatrix} \quad (3)$$

$$\varphi_l = \frac{2\pi d_l}{\lambda} n_l cos\theta_l \quad (4)$$

Where, l = 1,2. Then, we can obtain the reflection coefficient using the matrix elements of equation (3) as follow:-

$$r = \frac{f_{21}}{f_{11}}, R = |r^2| \quad (5)$$

Then, the transmittance can be obtained :-

$$T = |t|^2 = \left|\frac{1}{f_{11}}\right|^2 \quad (6)$$

$$A=1-(T+R) \quad (7)$$

Results and discussion:

In this section, we will discuss our simulation results. As figure (1) shows one -dimension semiconductor photonic crystal, where it comprises of two semiconductor materials repeated for N periods. The first layer is from ZnSe material with refractive index n_l=2.6, and thickness d_l, while the second layer is from TiN with refractive index n_2=1.288 and thickness d_2, at $d_1 + d_2 = \lambda/4$ substrates on glass, at the normal incident of light waves $\theta = 0^0$, and the number of periods of N=15 period. Figure (2) illustrates the transmittance spectrum versus the wavelength at the normal incidence of incident light waves. We have observed a photonic bandgap (PBG) in the ultraviolet (UV) regions with a width of 63.8nm, where the right edge is at 260.6nm, and the left edge is at 196.8nm. The transmission of UV light waves through this region is zero percent due to the Bragg scattering, which causes destructive interference between the incident light waves [28]. Now, we will show the effect of some parameters on the PBG region to control and engineering it.

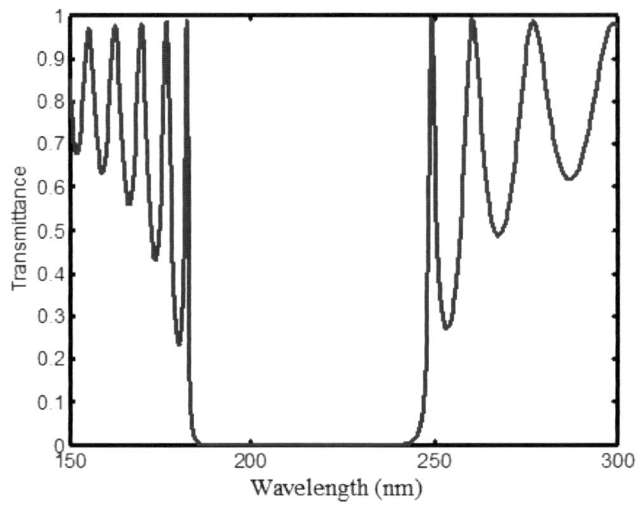

Figure 2: transmittance spectra of one-dimensional semiconductor photonic crystal consists of two materials ZnSe, and TiN at $\theta = 0^0$.

Figure (3) shows the effect of changing the thickness of the lattice parameter $\Lambda = d_1 + d_2$ based on the quarter wave stack, which is a multilayer of films with high- and low-refractive-indices, each with a thickness defined by:

$$\boldsymbol{d_i} = (\frac{\mathbf{1}}{\boldsymbol{n_i}})(\frac{\boldsymbol{\lambda_o}}{\mathbf{4}}) \qquad \textbf{(8)}$$

Wher d_i is the thickness of the layers i, n_i are the refractive index of layers, λ_o is the central wavelength of the reflection band. If we measure the reflectance of such a quarterwave stack as a function of the incident light wavelength (assuming a beam perpendicular to the surface), we will find a large reflectance band centered around λ_o. We can notice that the photonic bandgap shifted to a higher wavelength region, and a new two more sharp bandgaps.

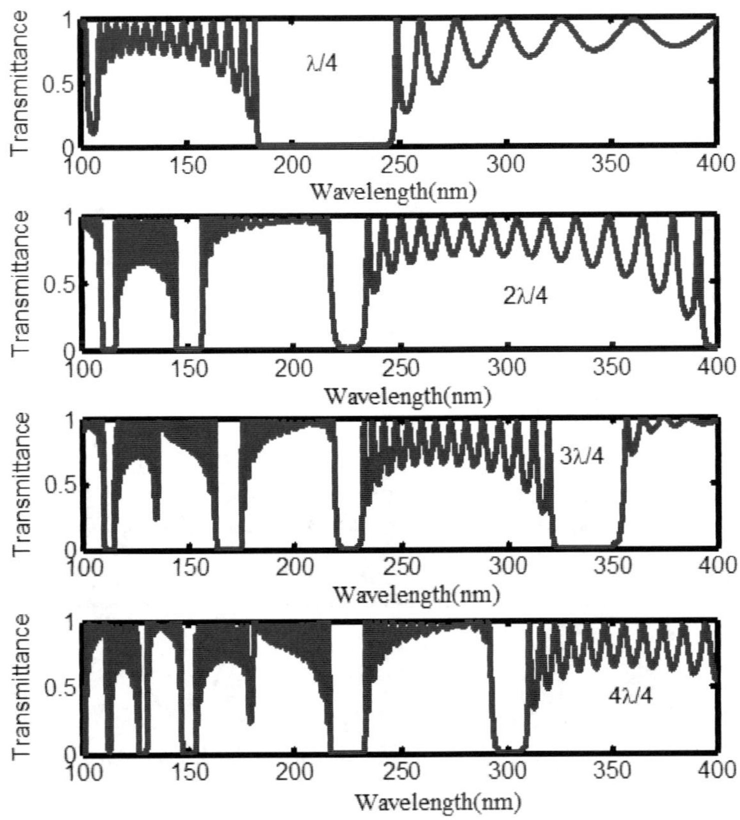

Figure 3: Studying the effect of changing the thickness of periodicity on the transmittance.

The first one appeared at a wavelength from 220.3nm to 231.9nm, the second bandgap at wavelengths from 144.7nm to 156.4nm, and the third one from 109.8 up to 115.9nm, as shown in figure (3). At $\Lambda = 3(\lambda/4)$, The band gap's width increases, and the number of band gaps also increases, reaching four bandgaps in the ultraviolet region. At $\Lambda = \lambda$, another new bandgap appeared in the UV region, raising the number to five band gaps, and the first bandgap shifted to a higher wavelength region. As we mentioned above, the thickness of ZnSe material in our proposed structure is an important parameter that could prevent a large amount of UV radiation due to the increase in the width of the bandgap and the appearance of new ones.

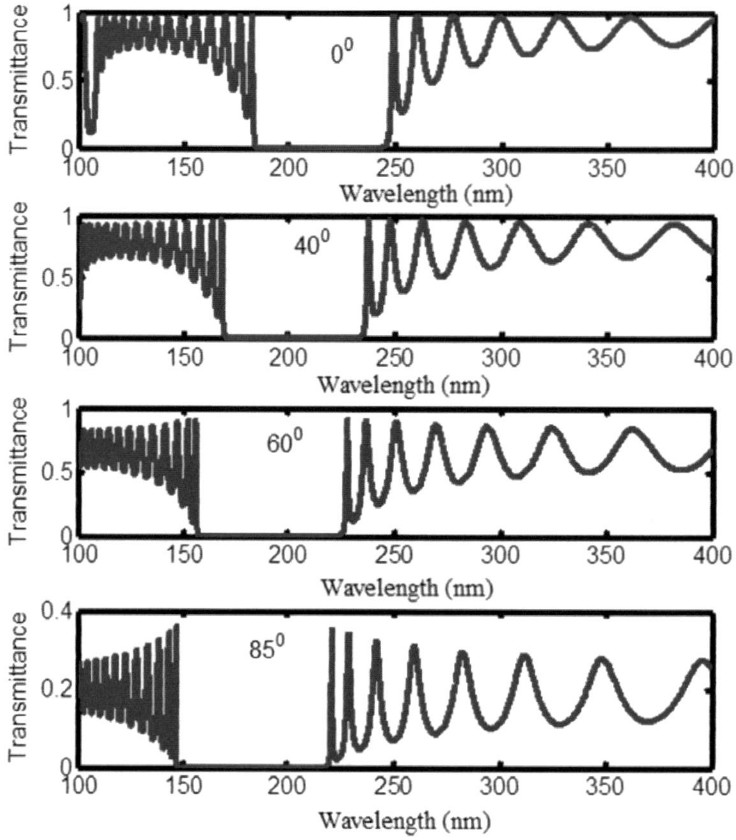

Figure.4: Transmittance spectra of the proposed structure with changing the incident angle.

In figure (4), we study the effect of changing the incident angle on our proposed structure and the PBG region, which is vital for preventing the harmful UV light waves. To study this effect, we fixed our structure thickness at $\Lambda = \lambda/4$ and, the number of periods N=15, In the case of the normal incidence $\theta = 0^0$, The bandgap width is 63.8nm. By increasing the incident angle to $\theta = 40^0$, we notice that the photonic bandgap region shifted to a lower wavelength region, increasing its width to be 66.3nm. At $\theta = 60^0$ the bandgap region has more shift to lower wavelength region, and the width increases to be 70.3nm. Finally, at $\theta = 85^0$the bandgap width reaches to be 83.7nm and shift more to a lower wavelength region. From these results, we can say that our proposed structure is beneficial in several optical filter applications such as the smart window due to the width of the photonic bandgap and the ability to controlling it.

Conclusion:

we performed a theoretical study on a one-dimensional photonic crystal consisting of two semiconductor material ZnSe and TiN, using the transfer matrix method. We investigated the effect of varying our structure parameters' such as thickness and incident angle on the transmittance spectra. Increasing Thickness cause increasing in the photonic bandgap number. Also, as the incident angle increases, a remarkable shift in the photonic bandgap happened. Furthermore, we can say that our proposal structure is instrumental in many optical applications, such as smart windows and optical filters.

References

[1] The American Cancer Society medical and editorial content team (2019) (www.cancer.org/cancer/acs-medical-content-and-news-staff.html) .

[2] Yablonovitch, E. (1987). Inhibited spontaneous emission in solid-state physics and electronics. Phys. Rev. Lett. **58**, 2059.

[3] John, S. (1987). Strong localization of photons in certain disordered dielectric superlattices. Phys. Rev. Lett. **58**, 2486.

[4] Aly, A.H., & Sayed, F.A. (2020). THz cutoff frequency and multifunction Ti2Ba2Ca2Cu3O10/GaAs photonic bandgap materials. Int.J. Mod. Phy. B, 34(10), 2050091.

[5] Aly, A.H., Mohamed, D., Mohaseb, M.A., Abd El-Gawad, N. S., & Trabelsi, Y. (2020). Biophotonic sensor for the detection of creatinine concentration in blood serum based on 1D photonic crystal. *RSC Advances*, 10, 31765 – 31772.

[6] Kabir, M.A., Hassan, M.M., Ahmed, K., M.S.Mani Rajan, Aly, A. H., Hossain, N., & Bikash Paul, B., K. (2020). Novel spider web photonic crystal fiber for robust mode transmission applications with supporting orbital angular momentum transmission property. *Opt. Quant. Electron.*, 52:331.

[7] Aly, A.H., Mohamed, D., & Mohaseb, M.A. (2020). Metamaterial Control of Hybrid Multifunctional High-Tc Superconducting Photonic Crystals for 1D Quasi-periodic Structure Potential Applications. Materials Research, 23(3): e20190695.

[8] Zaky, Z.A., & Aly, A. H. (2020). Theoretical Study of a Tunable Low-Temperature Photonic Crystal Sensor Using Dielectric-Superconductor Nanocomposite Layers. J Supercond Nov Magn., 33:2983–2990, .

[9] Aly, A. H., Mohamed, D., & Mohaseb, M.A. (2019). Theoretical and simulation study in defective semiconductor layer that incorporated with superconducting-dielectric photonic crystal. Int.J. Mod. Phy. B, 33(32), 1950397.

[10] Aly, A.H., Ameen, A.A., & Vigneswaran, D. (2019) . Superconductor Nanometallic Photonic Crystals as a Novel Smart Window for Low-Temperature Applications. J. Supercond Nov. Magn., 32(2), 191–197.

[11] Trabelsi, Y., Ali, N.B., Aly, A. H., & M. kanzari M., (2020). Tunable high Tc superconducting photonic band gap resonators based on hybrid quasi-periodic multilayered stacks. Physica C, 576, 1353706.

[12] Joannopoulos, J.D., Meade, M.R.D., & Winn, J.N. (2008). Photonic crystals: molding the flow of light. Princeton University presses, Princeton .

[13] Nouman, W.M., Abd El-Ghany, A. E.-S., Sallam, A. S., Dawood, A.B., & Aly, A.H. (2020). Biophotonic sensor for rapid detection of brain lesions using 1D photonic crystal. Opt Quant Electron., 52, 287.

[14] Aly, A. H., Mohamed, D., Elsayed H., & Mehaney, A. (2018). Fano-resonance by means of the one-dimensional superconductor photonic crystals. J. Supercond. Nov. Magn., 31, 3827–3833.

[15] Aly, A. H., & Mohamed, D. (2018). The Optical Properties of Metamaterial-Superconductor Photonic Band Gap With/Without Defect Layer. J Supercond. Nov Magn, 32(7), 1897–1902.

[16] Aly, A.H., Abdel Ghany, S.E.-S., B.M.Kamal, B.M., & Vigneswaran, D. (2020). Theoretical studies of hybrid multifunctional YaBa2Cu3O7 photonic crystals within visible and infra-red regions. Ceramics International, 46.1, 365-369.

[17] Aly, A. H., Ryu, S.W., Hsu, H.T., & Wu , A.J. (2009). THz transmittance in one-dimensional superconducting nanomaterial-dielectric superlattice. Materials Chemistry and Physics 113, 382-384.

[18] Kazuaki, S. (2005). Optical properties of photonic crystals. Springer, Berlin. 0342-4111

[19] Göde, F., Gümüş, C. & Zor, M. (2007). Investigations on the physical properties of the polycrystalline ZnS thin films deposited by the chemical bath deposition method. Journal of Crystal Growt , 299, 136–141.

[20] Grebe, G., Roussos, G. & Schulz, H. J. (1976). Cr^{2+} excitation levels in ZnSe and ZnS. *Journal of Physics C: Solid State Physics*, 9, 4511–4516.

[21] Hasnat, A., & Podder, J. (2012). Effect of Annealing Temperature on Structural, Optical and Electrical Properties of Pure CdS Thin Films Deposited by Spray Pyrolysis Technique," *Advances in Materials Physics and Chemistry*, 2(4),226-231

[22] Rumberg, A., Sommerhalter.C., Toplak.M., Jager-Waldau.A., MC Lux-Steiner, M.C. (2000). ZnSe thin films grown by chemical vapour deposition for application as buffer layer in CIGSS solar cells.Thin solid Films, 361-362, 172-176.

[23] Chu,T.L., Chu,S.S., J. Britt,J., C. Ferekides,C., Wu,C.Q. (1992). Films and junctions of cadmium zinc telluride. J. Appl. Phys, 71,5635

[24] Venkata Subbaiah,Y.P., Prathap,P., Devika, M., Ramakrishna Reddy,K.T. (2005). Close-spaced evaporated ZnSe films: Preparation and characterization. Physica B 365 ,240-246.

[25] www.Refractiveindex.Info

[26] www.filmetrics.com

[27] Yeh, P. (1988). Optical Waves in Layered Media. Wiley, New York.

[28] Mulder, H.K., Ymeti,A., Subramaniam,V., Kanger,J.S. (2012). Size-selective detection in integrated optical interferometric biosensors .Opt Express, 20,20934–20950.

Truncated Triangular Silver Nanoparticles using low concentration of CTAB

Hagar M. Saleh*, T. A. El-Brolossy, T. Abdallah, H. Talaat
Physics Department, Faculty of Science, Ain Shams University, Cairo, Egypt.

* E-mail: hagar.mustafa @sci.asu.edu.eg

Abstract. Triangular silver nanoparticles have special properties which increase their importance in chemistry, biology, physical and medical applications. This work concerns with fabrication of truncated nanotriangles, using a simple chemical method. The seed mediated growth method that is usually used to synthesize nanorods, is modified to create nanotriangles. By using this modified method, we successfully obtained silver truncated nanotriangles by adding low concentration of cetyltrimethylammonium bromide (CTAB) into seed solution. UV-visible spectroscopy (UV-Vis) was used to characterize the surface plasmon resonance of the formed nanoparticles. Also, the geometrical properties of nanoparticles were characterized by transmission electron microscopy (TEM) and scanning tunneling microscopy (STM). The average edge length of truncated nanotriangle is affected by varying the amount of seed added to the growth solution; it varies from 56 nm up to 75 nm according to the amount of seed used. The degree of truncation (T) for the produced nanotriangle edges has been estimated. Furthermore, the effect of CTAB concentration as well as the speed in which $NaBH_4$ is added to the seed solution on the formed nanoparticles was investigated.

Keywords: Nanotriangles, degree of truncation, CTAB, surface plasmon resonance.

1. Introduction

In last decades, there are great interest in nanomaterials, particularly plasmonic nanoparticles because of their properties and applications in various fields, specially biosensor ,nanoelectronics and nanophotonics [1, 2]. Plasmonics which are defined through the resonance of free electrons of metal nanoparticles and their quantum confinement lead to localized surface Plasmon resonance (LSPR). Triangular silver nanoparticles (TSN) as anisotropic nanoparticles have stronger plasmon resonance than nanorods and nanospheres. Consequently, they have a fundamental role in bioassay applications [3] as well as they are a key factor in surface enhanced spectroscopies [4]. Despite silver is of less stability than gold, LSPR of silver nanomaterials have superior sensitivity to local dielectric constant changes than gold. Also, enhancement factor and plasmonic property of Ag are better than other metals [5]. The optical properties of triangles nanoparticles with truncated edges have strong effect on the performance of LSPR sensor [6].

There are a various methods for synthesis of triangular silver nanoparticles such as photo-induced procedure and chemical reduction [7-10]. One of the most common chemical reduction methods is seed mediated growth method [11]. It is deemed basic protocol of synthesis anisotropic silver nanoparticles based on reduction of silver ions in the existence of CTAB and $NaBH_4$. CTAB is usually used to induce the orientation growth of seeds to form anisotropic nanoparticles [12]. It plays an important role in determining the morphology and size of the nanoparticles [13]. In spite of the role of $NaBH_4$ as strong reducing agent , it can increase stabilization ,uniformity and monodispersity of silver nanoparticles [14]. Also, Jill E. Millstone et al. [15] studied the effect of $NaBH_4$ concentration on nanoprism thickness and have shown change of optical spectra of the samples. Efforts are still needed to optimize methods of preparation of anisotropic nanoparticles, which depend on nature of chemical reaction, concentrations, speed of stirring, temperature, pH value and reducing agent [16].

In this work, we synthesized truncated TSN using a modified seed mediated method. The effect of adding various amounts of seed solution of low CTAB concentration to the growth solution has been studied through surface plasmon resonance, TEM and STM. Also, impacts of increasing CTAB concentration in the seed solution as well as the speed with which $NaBH_4$ is added to it on the surface plasmon resonance of created nanoparticles have been investigated.

2. Experimental

The procedure of preparation followed that of seed mediated synthesis by C.R. Rekha et al. at room temperature [17]. This method is based on reduction of silver ions by ascorbic acid as weak reducing agent. Silver nitrate is used as precursor for the solutions, as it is dissolved into the deionized water; the silver ions (Ag^+) are produced. Also, $NaBH_4$ is used as a strong reducing agent to create free metallic silver (Ag^o) atoms by the reduction. However, CTAB is considered as capping agent in growth solution.

In our modified method, the seed solutions were prepared by mixing 20 mL of 0.01 M $AgNO_3$ and 80 μL of CTAB of molar concentration 0.034 M (lower than that used by C.R. Rekha et al. [17]) under slow stirring. 0.6 mL of 0.01 M $NaBH_4$ ice cooled aqueous solution was added all at once. With two minutes stirring the color was bright yellow. The 1 h aged seed solution added to the growth solution which prepared as a stated in reference [17]. Different amounts of seed colloidal solution (0.5, 0.25, 0.125, 0.1, 0.6, 0.3 mL) were added. The color of the colloid solution changed within 3 minutes depending upon the concentration of the seed colloidal solution added. The purification of the silver nanotriangles was carried out by centrifuging at 6500 rpm for 30 min.

UV–Vis spectrometer (Ocean optics USA) was used to record optical absorption of surface plasmon resonance of the formed TSN colloidal solution. Morphology, size distribution and aggregation of the TSN were determined using transmission electron microscope (TEM, JEM -1200EX) and UHV Omicron scanning tunneling microscope (STM).

3. Results and Discussion

3.1 The influence of seed amount

Figure 1 shows the UV-Vis absorption spectra of synthesized TSN colloids using different amount of seed solution as indicated in the figure. The spectra of the TSN showed the three apparent predominant peaks related to different modes of plasmon excitation. According to Mie theory, the surface plasmon bands of plasmonic nanoparticles are due to dipole and quadrupole modes [18]. At the largest seed amount (0.5 mL), a strong band located at 490 nm is ascribed to the in-plane dipole while the band at 390 nm is attributed to out-of- plane quadrupole plasmon resonance. As shown in the figure, the broad peaks around 420 which may include the in-plane quadrupole and out-of-plane dipole becomes more apparent at small amount of seed solution. Also, decreasing the amount of seed solution causes these absorption bands to shift toward longer wavelengths (visibly demonstrated by varying color of solutions). The highest energy resonance band at about 340 nm which is assigned to the out of plane quadrupole resonance, is nearly invariable. Also, the full width at half maximum (FWHM) increases with decreasing the seed amount, varying from 133.47± 0.81 nm at the largest seed amount to be 235.52 ± 0.84 nm at the least amount. This red shift and broadening of FWHM can be ascribed to the increased charge separation during Plasmon oscillation as enabled by the increase in size. It is cleared that by changing the amount of silver seeds added to silver ions ,the size of the nanoparticles can be controlled [19].

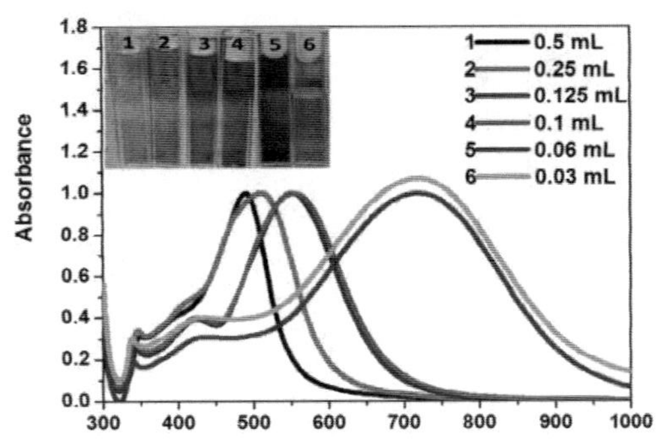

Figure 1 Color variations and UV–Vis spectra of TSN prepared with different amounts of seed solution.

IOP Conf. Series: Materials Science and Engineering **956** (2020) 012011 doi:10.1088/1757-899X/956/1/012011

3.2 TEM and STM measurements

TEM and STM measurements display the uniformity and structural properties of the prepared TSN. Figure 2 shows two chosen TEM images of TSN prepared using 0.25 mL (a) and 0.03 mL (b) seed solution. The images display the creation of triangular silver nanoparticles with truncated edges. This truncation in the edges is formed because of the oxidation of silver nanoparticles during the preparation [20]. Figure 3 shows STM image that confirm the truncation of the prepared triangle silver nanoparticles.

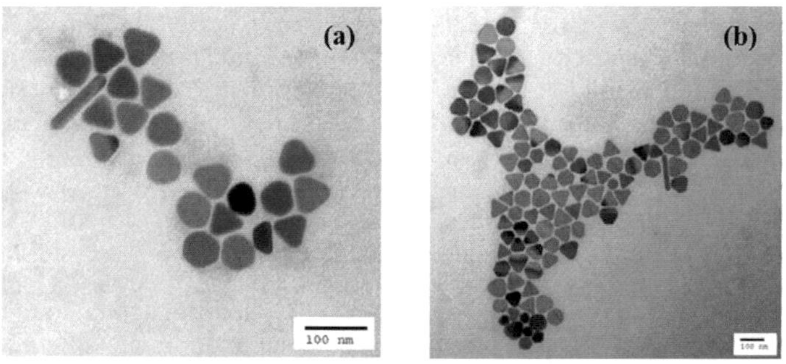

Figure 2 TEM images of TSN prepared at different amounts of seed solution (a) 0.25 mL and (b) 0.03 mL.

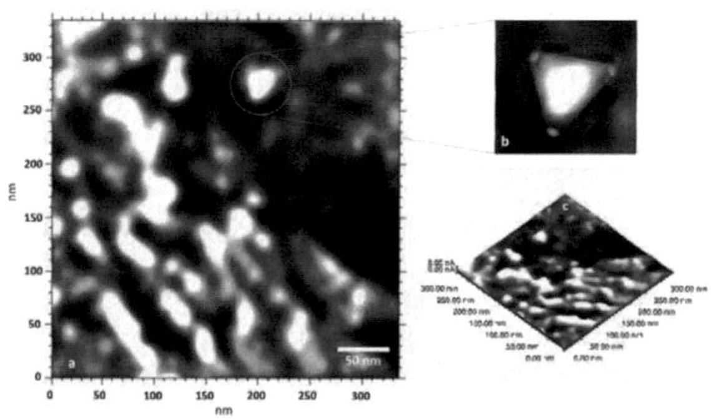

Figure 3 STM image of TSN at seed amount 0.03 mL.

It is observed that the TSN have edges with different degrees of truncation. The average degree of truncation (T) can be calculated using TEM images according to the relation T = b/a, considering the particle is composed by an inner triangle of side length "b" included into a parent triangle of side length "a" [9]. Clearly, a perfect triangle shape of the particle will be at T = 1, while T = 0 is for circular shape. Histogram of Figure 4 shows the average degree of truncation is 0.48 ± 0.0067 for sample of larger seed amount. When the particle size increased for sample of lower seed amount, the degree of truncation decreased to 0.41 ± 0.0068.

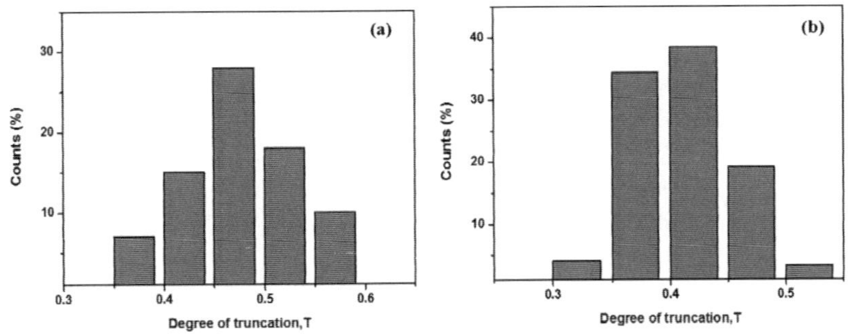

Figure 4 Histogram of degree of truncation distribution at (a) 0.25 mL and (b) 0.03 mL of seed amount.

Side length of TSN is an important parameter affecting absorption efficiency and resonance frequencies of dipole and quadrupole bands [21]. Using up to 200 particles of nanotriangles, it was found that the average side length is 56 nm with standard deviation ± 9.4 nm for sample of 0.25 mL of seed amount. The side length increases to 75 nm with standard deviation ± 12.3 nm for sample of 0.03mL seed amount as illustrated in Figure 5.

TEM images disclose the presence of other shapes of nanoparticles such as spheres, hexagons, rods, and others. Histogram of Figure 6 shows that most of the particles are nanotriangles and the minimum number of particles is nanorods. The nanotriangles occupy nearly 70% of nanoparticles in the colloids. The appearance of nano-hexagons may represent incomplete transformation to truncated nanotriangles.

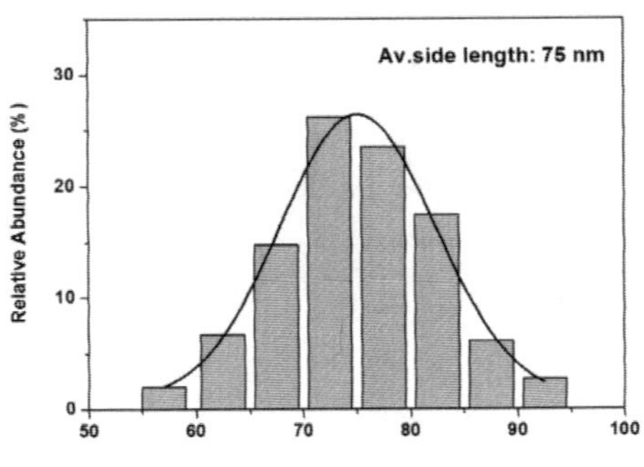

Figure 5 Side length distribution of the prepared TSN at 0.03 mL of seed amount.

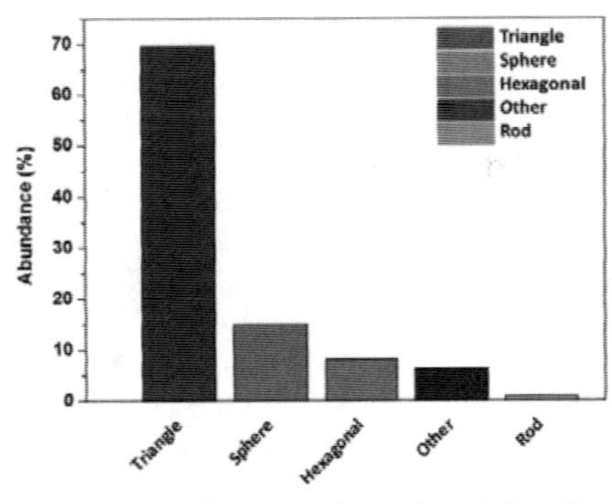

Figure 6 Histogram shows the relative abundance of different shapes using TEM images.

3.3 *The influence of CTAB in seed solution*

Using CTAB in seed as capping surfactant contributes to improve the regularity of the particles [22]. Also, the formed capping surface helps to prevent the aggregation of seeds which means that CTAB can be used as stabilizer. In this experiment we study the effect of using various concentration of CTAB in the seed solution.

Figure 7 shows UV-Vis spectra of TSN prepared with different total concentrations of CTAB in the seed solution (as indicated in the figure). These values are larger than the total concentration of CTAB (0.13 mM) used before. The amount of the seed solution added to the growth was fixed at 0.25 mL. It is observed that, the intensity of the in-plane dipole resonance band slightly decreases with irregular shift of its frequency with increasing the CTAB concentration. Also, increasing CTAB concentration has no significant effect on other resonance bands of TSN. This indicates that increasing CTAB concentrations is not influential factor which agrees with the results in reference [23].

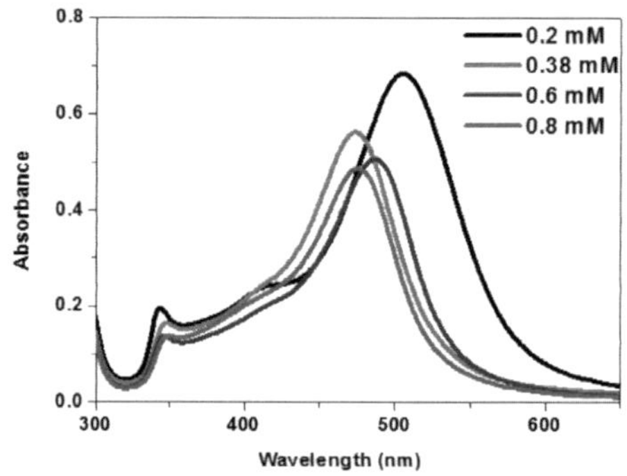

Figure 7 UV-Vis spectra of TSN prepared with different concentrations of CTAB.

3.4 The influence of sodium borohydride ($NaBH_4$)

How to add $NaBH_4$ is that significant? In this experiment we slowed down the speed of adding $NaBH_4$ to the seed solution, where 0.6 mL of 0.01 M $NaBH_4$ was added dropwise to the seed solution at rate of 0.018 mL for a total of 30 second with slow stirring. Figure 8 displays the absorption spectra of TNS prepared using three different amount of such seed solution (as indicated) of fixed CTAB concentration (0.13 mM). For fair comparison with the above results of UV-Vis spectra of TSN prepared via addition of $NaBH_4$ to the seed solution all at one shot, the three used amounts of the new seed solution in this case are similar to those used before with the same concentration of CTAB. With decreasing seed amount, the spectra show a larger red shift of the in-plane dipole band relative to that obtained for corresponding samples prepared by speedy addition of $NaBH_4$. This indicates that the particle size can be controlled by controlling the addition rate of $NaBH_4$ to the seed solution. Also, the intensity of in-plane quadrupole mode is significantly larger than that observed in the corresponding samples shown in figure 1, with a weak shoulder representing the out-of-plane dipole [10]. These observations can be attributed to the increase in thickness of nanotriangles with varying of charge density distribution [24]. These results imply that the speed of the addition of $NaBH_4$ significantly affects the particle size, thickness, and resonance modes of nanoparticles.

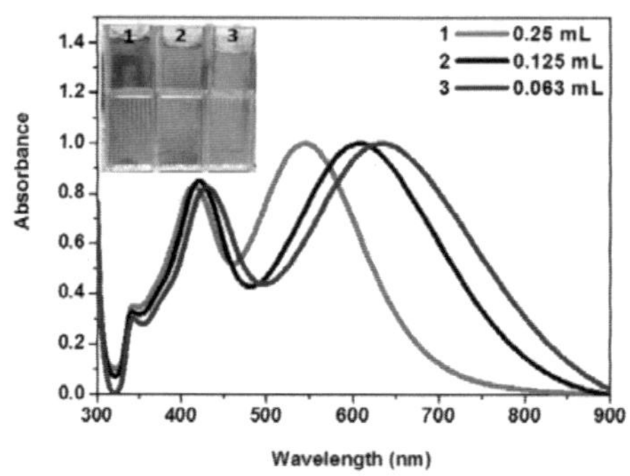

Figure 8 UV-Vis spectra and color variation of TSN prepared by addition of $NaBH_4$ drop by drop.

4. Conclusion

It is concluded that truncated triangle silver nanoparticles can be formed with seed mediate growth at low concentration of CTAB and low rate of stirring. Different degrees of truncation provide key control over many of the properties of nanoparticles and their applications. Also, rate of stirring and speed of adding $NaBH_4$ are effective parameters in tuning and enhancement of the surface plasmon resonance peaks of triangular nanoparticles which broaden their applications. Our results suggest that controlling the preparation conditions produces triangular nanoparticles appropriate to be used in in various fields like surface enhance Raman scattering and as sensors in the environmental applications. Furthermore, the stability of the geometrical shape with tunable plasmonic properties suggests that truncated nanotriangles are a candidate to play an important role in biomedical imaging.

Reference

[1] Diantoro M, Suprayogi T, Sa'adah U, Mufti N, Fuad A, Hidayat A and Nur H 2018 Silver Nanoparticles: Fabrication, Characterization and Applications 233
[2] Liu S, Tao H, Zeng L, Liu Q, Xu Z, Liu Q and Luo J-L 2017 Journal of the American chemical society **139** (6) 2160-2163
[3] Kelly J, Keegan G and Brennan-Fournet M 2012 Acta Physica Polonica A **122** (2) 337-345
[4] Wu C, Chen E and Wei J 2016 Colloids and Surfaces A: Physicochemical and Engineering Aspects **506** 450-456
[5] Yguerabide J andYguerabide E E 1998 Analytical biochemistry **262** (2) 157-176
[6] Ma W, Yao J, Yang H, Liu J, Li F, Hilton J and Lin Q 2009 Optics Express **17** (17) 14967-14976
[7] Song H, Zhang H, Sun Z, Ren Z, Yang X and Wang Q 2019 AIP Advances **9** (8) 085307
[8] Zannotti M, Rossi A and Giovannetti R 2020 Coatings **10** (3) 288
[9] Chen S andCarroll D L 2002 Nano letters **2** (9) 1003-1007
[10] Jin R, Cao Y, Mirkin C A, Kelly K L, Schatz G C and Zheng J 2001 science **294** (5548) 1901-1903
[11] Zhong Y, Liang G, Jin W, Jian Z, Wu Z, Chen Q, Cai Y and Zhang W 2018 RSC advances **8** (51) 28934-28943
[12] Zhang W, Hu G, Zhang W, Qiao X, Wu K, Chen Q and Cai Y 2014 Physica E: Low-dimensional Systems and Nanostructures **64** 211-217
[13] Törnblom M andHenriksson U 1997 The Journal of Physical Chemistry B **101** (31) 6028-6035
[14] Kruis F E, Fissan H and Rellinghaus B 2000 Materials Science and Engineering: B **69** 329-334
[15] Millstone J E, Hurst S J, Métraux G S, Cutler J I and Mirkin C A 2009 small **5** (6) 646-664
[16] Khodashenas B andGhorbani H R 2019 Arabian Journal of Chemistry **12** (8) 1823-1838
[17] Rekha C, Nayar V and Gopchandran K 2018 Journal of Science: Advanced Materials and Devices **3** (2) 196-205
[18] Zhang J, Li S, Wu J, Schatz G C and Mirkin C A 2009 Angewandte Chemie **121** (42) 7927-7931
[19] Raza M A, Kanwal Z, Rauf A, Sabri A N, Riaz S and Naseem S 2016 Nanomaterials **6** (4) 74
[20] Yang H, Owiti E O, Jiang X, Li S, Liu P and Sun X 2017 Nanoscale research letters **12** (1) 1-6
[21] Chen Q, Qi H, Ren Y-T, Sun J-P and Ruan L-M 2017 Aip Advances **7** (6) 065115
[22] Nikoobakht B andEl-Sayed M A 2003 Chemistry of Materials **15** (10) 1957-1962
[23] Hu G, Liang G, Zhang W, Jin W, Zhang Y, Chen Q, Cai Y and Zhang W 2018 Journal of materials science **53** (7) 4768-4780
[24] Dreaden E C, Near R D, Abdallah T, Talaat M H and El-Sayed M A 2011 Applied Physics Letters **98** (18) 183115

IOP Conf. Series: Materials Science and Engineering **956** (2020) 012012 doi:10.1088/1757-899X/956/1/012012

Helical Structures for Medical Applications

M Eldeeb[1], H Mattar[2], H Hatem[1], D Elmahdy[2], N Shanan[3], M Elwi[2], A Klingner[4], A Ramzy[2*]

[1] Mechatronics Engineering Department, Faculty of Engineering and Materials Science, German University in Cairo, New Cairo 11835, Egypt.
[2] Materials Engineering Department, Faculty of Engineering and Materials Science, German University in Cairo, New Cairo 11835, Egypt.
[3] Department of Pharmacology and Toxicology, Faculty of Pharmacy and Biotechnology, German University in Cairo, Egypt.
[4] Physics Department, Faculty of Engineering and Materials Science, German University in Cairo, New Cairo 11835, Egypt.
* Email:amna.ramzy@guc.edu.eg

Abstract. The latest advances in the field of microrobotics assisted medical experts in the treatment of some of human diseases especially intravascular diseases such as atherosclerosis. In this research, a phantom for the atherosclerotic plaque material is fabricated using Gelatine and Nano Hydroxyapatite in a catheter tube to mimic the condition in the human body in vitro. For plaque penetration a helical microrobot is designed to achieve the optimum plaque clearing efficiency. The robot is a remotely controlled micro helical magnetic based metallic robot which is inserted in the catheter to penetrate the phantom by rotational motion along its axis. The motion of the robot during mechanical grinding of the plaque phantom is simulated using MATLAB where the distance of the robot's journey inside the plaque phantom is plotted against the time consumed by the robot.

1. Introduction

Atherosclerosis is a disease in which plaque is created inside the arteries. Plaque consists of cholesterol, fats, calcium, and other substances that are deposited on the walls of the arteries. As time passes, the plaque becomes harder and accordingly the arteries become narrower. This decreases the flow of oxygen-rich blood all over the human body. Atherosclerosis is capable of causing stroke and heart disease.[1] Cardiovascular diseases (CVDs) are a common cause of death all around the world. In 2016, approximately 17.9 million people died because of CVD, and this represents 31% of worldwide deaths and 85% of these deaths are because of stroke and heart diseases.[2]

In 2005, Hansson, G.K. stated that atherosclerotic lesions (atheromata) consist of lipids, debris, cells, and tissue elements. Another component that forms the plaque is calcium phosphate hydroxyapatite.[3] In 1994, Demer *et al.* pointed out that in coronary arteries, when the plaque is built up the amount of calcium can be 10 to 80 folds. By using intravascular ultrasound, the calcified plaques are more than 80%. 90% of patients who have coronary artery disease had coronary calcification.[4]

In their research, Lihong Lao *et al.* studied the arterial plaque composition and simulated its behaviour by selective mineralization of hydrogel lumens to be able to perform ex-vivo experiments. Using polyacrylamide, the synthetic lumens were moulded by casting the acrylamide solutions into a transparent cylindrical mould to mimic the arterial plaque.[5] Another research that focused on the hard part of the plaque was concerned with how to tackle and remove the plaque using rotational ablation. In their model, a mixture of Hydroxyapatite and Polylactic acid was used and its resemblance to the calcified plaque was confirmed using Nano indentation.[6]

In 2018, Lee, S. *et al.* built up Magnetic Drilling Actuators, that were made of Ni and cobalt (Co) and were covered with titanium (Ti) or titanium oxide (TiO_2) for the microrobots to be biocompatible. To improve the performance of the drilling and overcome the torque and force of gravity, they used Nd permanent magnet.[7]

In 2010, Park, S. *et al.* noted that robots can be helpful in coronary interventions. They are willing to make a microrobot that should be able to move freely and recognize the target location which would serve as both an effective drilling tool and a thrombus collector.[8]

In this research paper, a phantom of the arterial plaque was fabricated inside a catheter tube using gelatin and hydroxyapatite. A magnetically controlled helical micro-robot was then introduced to the catheter and the penetration of the robot inside the plaque phantom was studied.

2. Materials and Methods

The experiments are divided into three main stages as follows;

2.1 Stage 1: Plaque Phantom Fabrication

Three mixtures of Gelatine/Hydroxyapatite (HA) were prepared in the following compositions: 5wt%, 10wt% and 15wt% solute in solvent (boiled distilled water); with a ratio of 60:40 HA to Gelatin.

The Gelatin was mixed with HA in boiled distilled water using a magnetic stirrer at 600±5 rpm for 15 minutes. The mixture is then inserted in a catheter tube of diameter 5 mm and plugged from both ends then left to cure for 1 hour at 5°C.

2.2 Stage II: Fabrication of Helical Microrobot

The helical microrobot is made of a copper wire of diameter 0.3 mm that is wound in a spring form with a pitch of 1.3 mm, a total length of 5 mm and a diameter of 1 mm. A 1 mm diameter permanent magnet supplied by Supermagnete, Gottmadingen, Germany is mounted on the robot as shown in figure 1 to provide the required magnetic properties. The magnet is mounted so that the magnetic moment of the head is perpendicular to the axis of the robot.

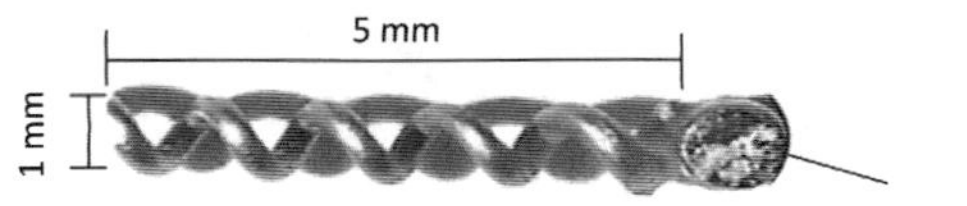

Figure 1. Helical microrobot.

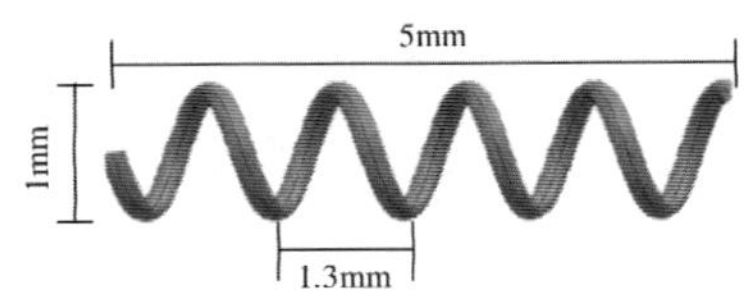

Figure 2. Helical microrobot by solidworks.

2.3 Stage III: Plaque penetration by helical microrobot

IOP Conf. Series: Materials Science and Engineering **956** (2020) 012012 doi:10.1088/1757-899X/956/1/012012

The penetration experiment takes place inside a special set-up called HeliMag shown in figure 3. The helical microrobot is steered and propelled under the effect of the magnetic fields. The magnetic fields are generated by two rotating permanent magnets. An Arduino control board is used to control and synchronize their motion. To minimize vibrations, the system is put on tuned damped upgradable smart table.

Figure 3. HeliMag System.

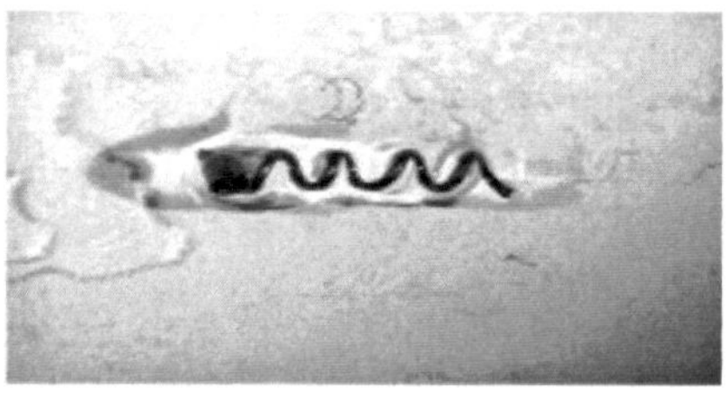

Figure 4. Helical microrobot under the Microscope.

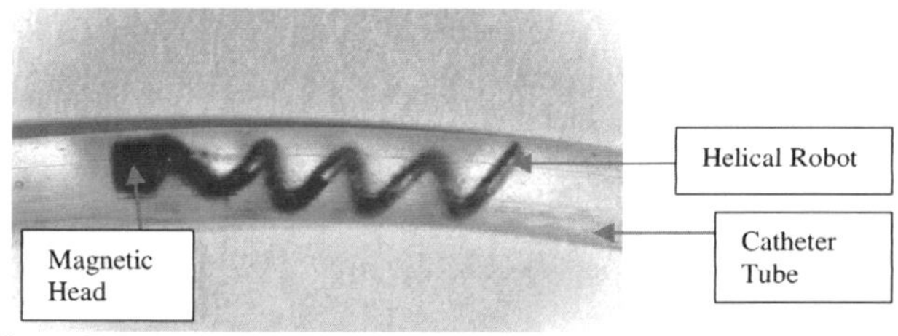

Figure 5. Robot inside catheter tube under Microscope.

2.4 Stage IV: Speed of the Robot & Rubbing Simulation

By using modelling equations, the speed of the helical robot and the size of the part removed from the plaque were calculated and then the penetration process was simulated and visualized using MATLAB. The plaque mass was modelled as a cylindrical shape of diameter 4 mm and length 4.5 mm.

First, the magnetic torque which is exerted on the robot was calculated by using

$$T_m = m \times B, \tag{1}$$

Where m is the magnetic moment and B is magnetic field. When the robot swims in liquid medium (water), the fritting torque (Tf) ≃ 0, therefore, the drag torque (Td)= Tm. From the equation of drag torque, the speed of the robot was measured.[9]

$$\begin{pmatrix} F_d \\ T_d \end{pmatrix} = \begin{pmatrix} A_{11} & A_{12} \\ A_{12} & A_{22} \end{pmatrix} \begin{pmatrix} U \\ \Omega \end{pmatrix}, \tag{2}$$

where U, Ω are the linear and angular velocity of the robot, and the parameters of the matrix are:

$$A_{11} = 2\pi nr \left(\frac{\zeta_\parallel \cos^2\theta + \zeta_\perp \sin^2\theta}{\sin\theta} \right), \tag{3}$$

$$A_{12} = 2\pi nr^2 (\zeta_\parallel - \zeta_\perp) \cos\theta, \tag{4}$$

$$A_{22} = 2\pi n r^3 \left(\frac{\zeta\perp \cos^2\theta + \zeta\| \sin^2\theta}{\sin\theta}\right), \quad (5)$$

Where n, r are the number of turns and radii of the helix, and $\zeta_{\|}$, $\zeta_{\perp}$ are the parallel and normal viscous drag coefficients.

$$\zeta_{\|} = \frac{4\pi\eta}{\ln\left(\frac{0.36\pi r}{b\sin\theta}\right)+0.5}, \quad (6)$$

$$\zeta_{\perp} = \frac{2\pi\eta}{\ln\left(\frac{0.36\pi r}{b\sin\theta}\right)}, \quad (7)$$

Where η is the fluid viscosity and b is the radii of tail.[9]

If the robot touches the plaque and does fitting, then its speed≃ 0 and T_d = A22*omega. Therefore, the size of the removing part from the plaque(δbc) is calculated using [10]

$$\delta_{bc} = \min\left(\frac{2\sqrt{|Tm|-|Td|}}{\sqrt{\pi\tau r}}, 0.5b\right), \quad (8)$$

3. Results and Analysis

The discussion of this research will be divided into two parts the first part is the experimental part where the plaque phantom penetration with the helical robot is presented. The second part is the simulation part where the speed and motion of the helical microrobot during rubbing inside the plaque will be visualized and discussed in comparison to the experimental results.

3.1 Plaque phantom penetration

The fabricated plaque is put inside the catheter tube produced. The plaque has a white high viscosity mass as shown in figure 6.

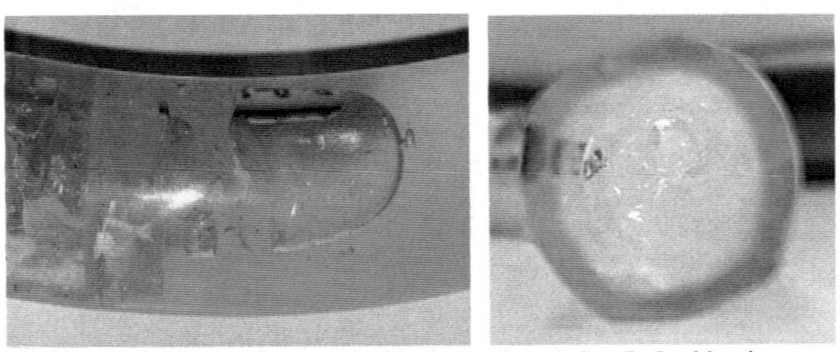

Figure 6. Plaque Phantom inside the catheter tube: *Left:* side view, *Right*: cross section.

By visual and physical inspection (applying pressure by the finger), the 5wt% plaque sample was soft yet holding its own shape and not flowing; elastic yet easily penetrated with small pressure. The 10wt% sample was harder than the 5wt% sample and even more elastic yet it was penetrated with enough pressure. The 15wt% sample was the hardest of them all resembling a soft rubbery texture and was the most elastic and could only be penetrated with the use of excessive force.

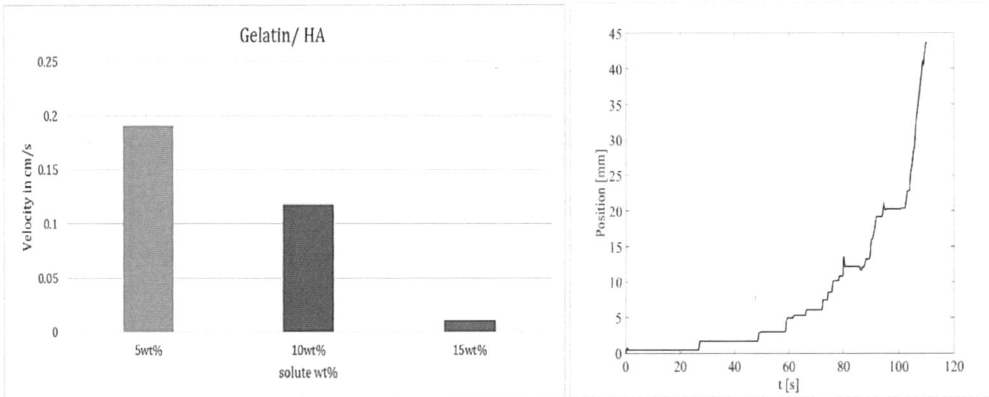

Figure 7. *Left:* Average Velocity of the robot with different concentrations of Gelatin/HA, *Right:* Detailed velocity Profile of the robot in the 10wt% concentration medium.

In the sample with 5wt% composition, the helical microrobot could easily cover an average distance of 0.191 cm in one second whereas by increasing the wt% of HA, the rigidity of the phantom plaque mass increased causing more resistance on the robot and consequently it covered a smaller distance (0.118 cm) in the same time.

A remarkable drop in the velocity of the robot could be noticed by further increasing the amount of HA to 15wt% where the velocity dropped by almost 10 fold from 0.118 to 0.011 cm/s.

It can be easily concluded that the harder the plaque phantom is, the slower the penetration velocity of the robot gets. This could be attributed to the presence of high frictional forces between the robot body and the abrasive HA content. This same results was also discussed by Chenghao Bi et al [11] where the authors demonstrated the ability of a magnetic microrobot to travel through wet conditions inside a murine colon. The results showed that when starting in water medium the robot reached a maximum speed of 2.2 mm/s while the speed of the robot inside 1% Tylose solution which resembles the ultrasonic hydrogel decreased to roughly 0.2 mm/s.

The motion of the helical microrobot throughout the whole journey of penetration in the 10wt% sample is detailed in figure 7 (right). From the graph, it can be noticed that the robot doesn't follow a constant course of motion where the velocity changes stepwise according to the position of the robot inside the plaque mass. During the penetration experiment, it was observed that the robot rotates along its axis to rub and remove a small plaque mass then it proceeds linearly afterwards matching another step on the graph. This nature of rubbing was also simulated by Dai et al [12] where the dynamic motion of a rubbing rotor was simulated and studied. It was found out that the rotor speed was not constant all over its rubbing motion and rather creates a whirl like orbits whose amplitude was directly proportional to the applied frequency level. This varying amplitude occurs due to the rubbing friction and reflects on the stability of the rotational speed of the rotor.

3.2 Rubbing motion simulation

The removing part of plaque δ_{bc}= 0.075mm was determined using equation (8). Then, the translation and rotation of the helical robot while moving in water was simulated. Then, the penetration process of the helical robot against the plaque was visualized.

When the robot first contacts the plaque, it rotates along its axis without linear motion. The rotational motion of the helical robot causes the rubbing and consequently the removal of the plaque at the contact area, this in turn gives the robot a free volume allowing it to move forward and further penetrate into the plaque mass. This process will be done all over again till the robot reaches the end of the tube. In the model, 1 pixel is equals to δ_{bc} and the total length of the tube is given by 15mm and the diameter 4.5mmAll dimensions were converted to pixel by dividing by δ_{bc}. In figure 8, the cross section of the catheter tube containing the robot is visualized in two different positions; a) water medium, b) plaque medium, c) is the complete side view of the tube showing both phases water (blue) and plaque (orange). The pixeled micro-helical robot is shown in d).

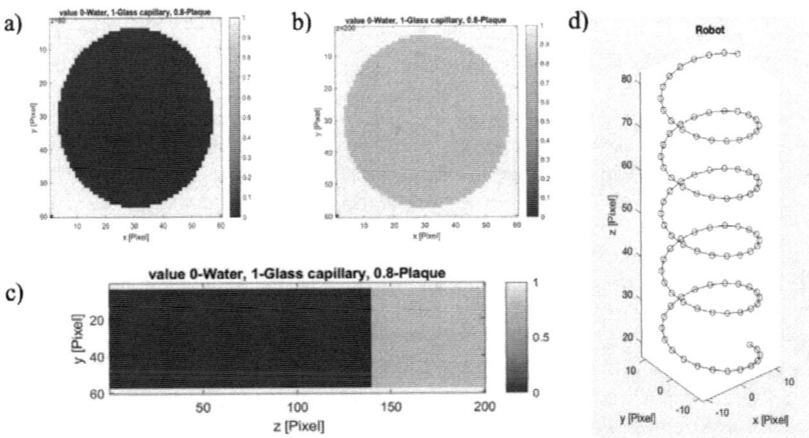

Figure 8. a) Cross section of the tube with water, b) Cross section of the tube with plaque, c) Side view of the tube, and d) Helical Microrobot.

In figure 9(a), the motion of the helical robot in the water medium (blue) is visualized where it can be noticed that the velocity of the robot is constant since there is low resistance exerted by the Newtonian fluid. The graph in figure 9(b) shows the volume of the plaque in mm^3 over time in seconds. As the robot is still in the water medium and it does not touch the plaque, the volume of the plaque remains constant. The model succeeds in visualizing the dynamic motion of the robot as the medium changes from water to plaque as shown in figure 9(c) where the velocity of the robot changes in stepwise accompanied by a decrease in the volume of the plaque as illustrated in figure 9(d).

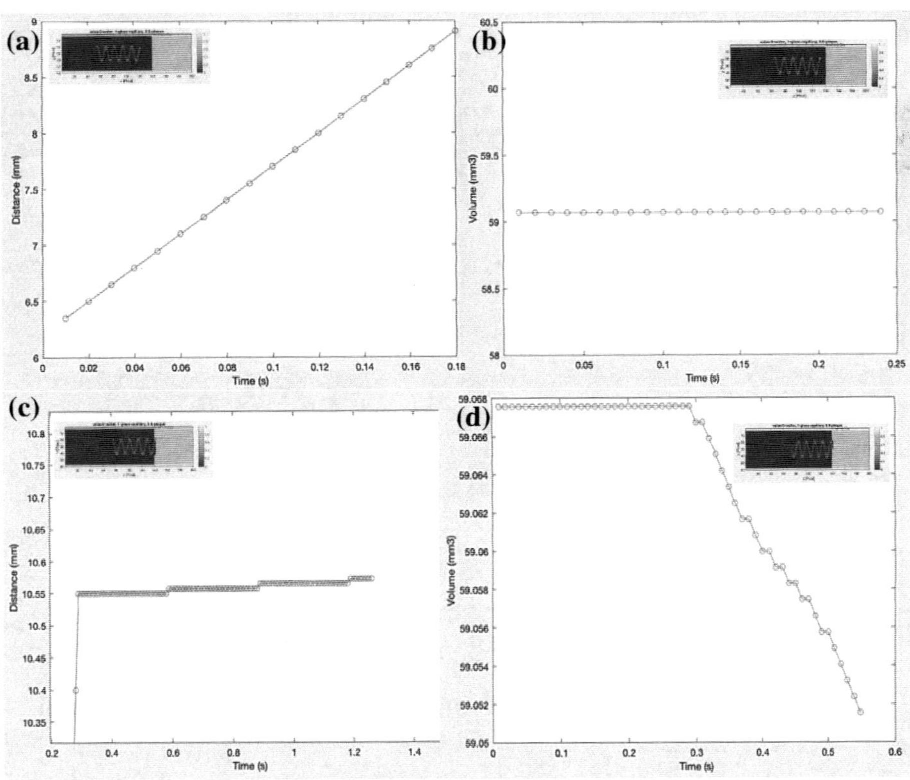

Figure 9. Position of the robot versus time (a,c), Volume of the plaque over time (b,d).

The simulation of the robot behavior was successfully simulated using the model presented in section 2.4. The same dynamic behavior of magnetic microrobots was also discussed by several researchers where Öpöz and Chen [13] used the FEM method to simulate the rubbing behavior of a single grain. The behavior of the grain was found out to pass by three main stages namely: rubbing, ploughing and cutting resembling the elastic deformation, elastic-plastic deformation and cutting zones respectively. The path of grain was simulated over the three stages where the researchers illustrated changes in the grain displacement path and the exerted forces on the grain related to the zone. The simulation results showed that the forces begin at high values (in our microrobot case is the phase where rubbing occurs without linear displacement) and then a subsequent decrease in the forces occurs due to the removal of material (in our case the linear penetration of the robot inside the plaque mass after volume decrease).

The same rubbing behavior was also discussed by Khalil et al. where the drag forces and the torque exerted on the magnetic micro-helical robot was simulated [10]. The results showed that the forces and torques induced during the rubbing action are fluctuating which cause the variation of the robot speed along the penetration path.

4. Conclusion

In this research paper, a plaque phantom was fabricated using a mixture of gelatine and Hydroxyapatite at 5wt%,10wt% and 15wt% to simulate the behaviour of the helical microrobot during rubbing motion *in vitro*. The motion of the robot inside the plaque phantom was captured, analysed and the average velocity at each concentration of HA was calculated. A correlation between the concentration of the HA and the robot velocity was deduced where maximum robot velocity was achieved at 5wt% hydroxyapatite. A detailed velocity curve of a selected concentration (10wt%) was then plotted. The rubbing motion behaviour of the helical microrobot was modelled then simulated and visualized using MATLAB.

The detailed velocity curve was then used to verify the simulated robot motion as seen in figure 10. The blue colour resembles water medium where there is no friction or resistance on the robot in the first 20 seconds. Starting second 23 the robot tip contacts the plaque (yellow colour) and friction occurs causing the slope of the curve to change. Full penetration occurs after t=100 s where the speed of the robot increases by exiting the resisting medium.

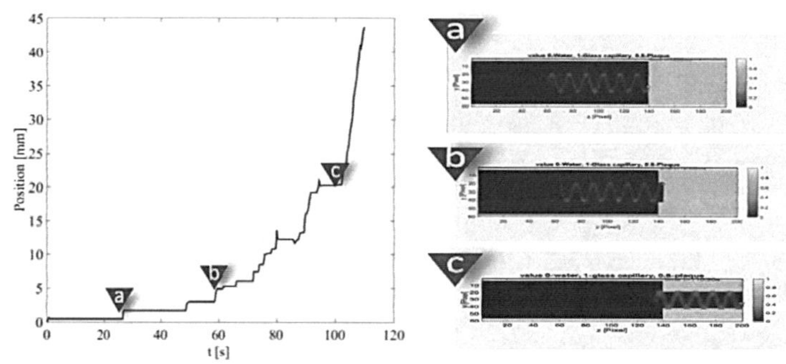

Figure 10. Position versus time of the robot motion passing through 5wt% of HA (a) The robot touched the plaque. (b) The robot started the penetration process. (c) The robot reached the end of the plaque and then moved to the water medium.

References

1. Arteriosclerosis / atherosclerosis Symptoms & causes. *Mayo Clinic* April 24, 2018 www.mayoclinic.org/diseases-conditions/arteriosclerosis-atherosclerosis/symptoms-causes/syc-20350569

2. Cardiovascular diseases (CVDs). *World Health Organization* May 17, 2017 www.who.int/en/news-room/fact-sheets/detail/cardiovascular-diseases-(cvds)

3. Hansson G 2005 Inflammation, Atherosclerosis, and Coronary Artery Disease. *New England Journal of Medicine* **352(16)**, 1685–1695.

4. Demer L, Watson K and Boström K 1994 Mechanism of calcification in atherosclerosis *Trends in cardiovascular medicine* **4(1).** 45-49.

5. Lao L, Robinson S, Peele B, Zhao H, Murray B, Min J, Mosadegh B, Dunham S and Shepherd R 2017 Selective Mineralization of Tough Hydrogel Lumens for Simulating Arterial Plaque. *Advanced Engineering Materials* **19.** 1-8.

6. Kim M, Kim H, Kim N, Yoon H and Ahn S 2011. A rotational ablation tool for calcified atherosclerotic plaque removal. *Biomedical microdevices* **13(6)**, 963–971.

7. Lee S, Lee S, Kim S, Yoon C, Park H, Kim J and Choi H 2018 Fabrication and Characterization of a Magnetic Drilling Actuator for Navigation in a Three-dimensional Phantom Vascular Network. *Scientific Reports* **8(1).** 1-9.

8. Park S, Cha K and Park J 2010 Development of Biomedical Microrobot for Intravascular Therapy. *International Journal of Advanced Robotic Systems* **7(1)**, 91-98.

9. Abbott J, Peyer K, Lagomarsino M, Zhang L, Dong L, Kaliakatsos I and Nelson B 2009 How Should Microrobots Swim? *The International Journal of Robotics Research*, **28(11-12)**, 1434–1447.

10. Khalil I, Tabak A, Sadek K, Mahdy D, Hamdi N and Sitti M 2017 Rubbing Against Blood Clots Using Helical Robots: Modeling and In Vitro Experimental Validation. *IEEE Robotics and Automation Letters*, **2(2)**, 927–934.

11. Bi C, Niedert E, Adam G, Lambert E, Solorio L, Goergen C and Cappelleri D 2019 Tumbling Magnetic Microrobots for Biomedical Applications. *2019 International Conference on Manipulation, Automation and Robotics at Small Scales (MARSS).* **(-)**, 1-6.

12. Dai X, Jin Z and Zhang X 2002 DYNAMIC BEHAVIOR OF THE FULL ROTOR/STOP RUBBING: NUMERICAL SIMULATION AND EXPERIMENTAL VERIFICATION. *Journal of Sound and Vibration*, **251(5)**, 807–822.

13. Öpöz T and Chen X 2010 An Investigation of the Rubbing and Ploughing in Single Grain Grinding using Finite Element Method. *In: 8th international Conference on Manufacturing Research.* **(-),** 256-261.

IOP Conf. Series: Materials Science and Engineering **956** (2020) 012013 doi:10.1088/1757-899X/956/1/012013

Blue, Red, and Green Emission from Chromium and Copper Metal Doped Lithium Borate Glass

I. Kashif*, A. Ratep

*Physics Department, Faculty of Science, Al-Azhar University, Nasr City, Cairo, Egypt.
Department of Physics, Faculty of women for Art, science & education, Ain Shams University, Heliopolis, Cairo, Egypt

Email: ismailkashif52@yahoo.com, Tel.01006609924 Fax: 24558035

Abstract The influence of copper metal on the structural and optical properties (absorption and emission) of the lithium borate glass samples containing chromium considered. The analysis of FTIR spectra measured, indicated the copper and chromium ions act as network former and modifier depend on the concentration of everyone in the sample content. The ratio between the BO_4 groups to the total boron groups (BO_4+BO_3) decreases with the copper content increases up to 2 mol % and then increase. The optical properties that the Dq/B decreases with the increase the copper and higher than 2.3 which indicates the glass samples containing Cr^{3+}ions in the high-field sites and produce conventional laser action. The emission spectra show the intensity emission bands at 440 nm, 540 nm, and 730 nm for relative strong blue, green and red bands respectively.

Keywords: Lithium borate glass; FTIR spectroscopy; optical properties (absorption and emission); blue, green and red emission

1. Introduction

Borate glasses have low melting point, high thermal stability, optical transparency, large non-linear optical properties, and good rare earth RE /transition metal (TM) ion solubility; this makes it suitable for potential applications in nonlinear optics.

The addition of alkali oxide to metaborate glasses improved its various technical and industrial applications. What's more, the borate glasses imparts semiconducting /conducting nature to an otherwise insulating material. Borate glasses doped with heavy metal ions or transition metals considered to have good nonlinear optical properties[1][2][3]. Transition metal ions contribute partially filled d-orbital to glass, and electro-optical behavior observed between ions in lower and higher oxidation state[4]

Glasses containing divalent cations such as Cu^{2+} have semiconducting semiconductor properties and glasses with a trivalent ion such as Cr^{3+} show luminescent and magnetic properties[5]. The anomalous behavior of the borate glass makes them of interest due to super structural units formed by BO_3 and BO_4 units.

These two supercooled structural units were formed either from the so-called supercooled or from various structural groups such as diporate, tetraborate, pentameric, boroxol rings, etc[4].

The presence of copper in the glass sample has many advantages as they are

(i) Potential materials due to their natural abundance.

(ii) Easiness of Cu oxidation (Cu^{2+} and Cu^{+} ions)[6].
(iii) May also exist as Cu^{0} ions.
(vi) Promising for photovoltaic energy conversion, display bright photoluminescence depending on the host.
(vii) The disintegration of little amounts of copper particles in glass networks makes the glasses colored and emphatically impacts the structural and optical properties[7].
Chromium ions are very effective and useful nucleation agents from the different transition metal ions as it exists in different valence states (Cr^{3+} and Cr^{6+}), Cr^{3+} ions occupy the octahedral sites because the $3d^3$ electronic configuration has maximum stabilization energy in the six-fold octahedral coordination[5][8]. Lesniewski et al.[9]contemplated the optical spectroscopic portrayal of calcium tetraborate glasses containing Cr^{3+}.
The luminescence excitation comprised of two broadband peaks corresponding to transitions from the $^4A_{2g}$ to the $^4T_{1g}$ and $^4T_{2g}$ levels of the octahedral coordinated Cr^{3+} ions.
Kaczmarek[10] studied $Li_2B_4O_7$ glasses containing Cr, Co, Eu, and Dy. They found that the sample containing Cr^{3+} ions are good optical quality, giving clear luminescence.
Stefan et al.[6] studied the effect of copper content on the structural and optical properties of zinc borate glasses and they found that the copper ions act as the network modifier and formed the NBOs in the examined glasses.
Many researchers[7][11][12][13] studied the effect of addition, copper in borate glass and they found that some glass samples have various characterizations reveal that highly transparent, thermally steady, amorphous, and homogeneous with precise elemental traces.
From our previous work[12][14] lithium tetraborate glass containing chromium metal has more thermal stability, high transmission and emission intensity than the glass with chromium oxide. The addition of copper metal to borate glass increases the glass homogeneity and can act as a filter to cut the light in the UV and infrared range. Also, increases the NBOs than CuO. And have higher transmission intensity.
The addition of copper or chromium metal on a borate glass system introduces a defect like the color center and wrong bond in the glass. In view of the above focal points, so in the present article, the point of this examination is fascinating to research the impact of the glass creation and the replacement of copper metal by chromium metal concentration on the spectroscopic properties, the stability of glass, the transmission and the emission properties of lithium metaborate glasses.

2. Experimental work

Glass samples composition 48 Li_2O–48 B_2O_3–(4-x) Cr metal–x Cu metal [x= 0, 1, 2 and 3] mol % synthesized. The melt quenching technique used in the prepared of glass samples. The pure chemicals mixed and placed in high temperature furnace (1100°C) in a porcelain crucible. The differential thermal analysis (DTA) a Shimadzu 50 Analyzer used to estimate the thermal parameters. The glass samples crushed to fine powder and examined by X-ray diffraction system, (Philips Analytical type PW3710).
The density of the glass samples determined using an Archimedes method. The FTIR absorption spectra of the glass samples obtained at room temperature in the range 2000– 400 cm^{-1} by an infrared spectrometer (type JASCO FTIR-4100, Japan) utilizing

the KBr plate method.
The structural groups of the glass samples determine by deconvoluting of the FTIR spectra using the winding program.
The optical absorption spectra obtained using a UV-Vis-NIR spectrophotometer (type JASCO, V-570) in wavelength range 190-2500 nm. The emission measurement utilizing (JASCO– FP-6300) Spectrofluorometer in the wavelength range (200– 800 nm).

3. Results and discussion

Figure 1 demonstrates the XRD glass samples. A broad diffraction peaks observed in all X-ray patterns of the glassy samples. This confirms the absence of any crystalline phase and indicates the amorphous state of the glass samples.

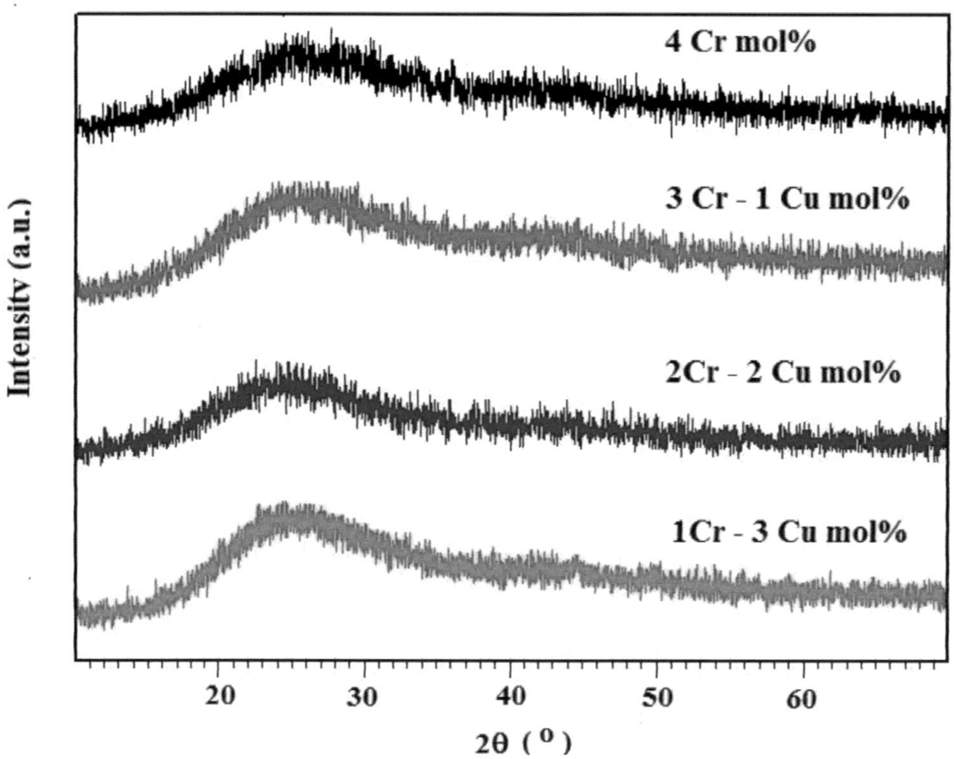

Figure 1 X-ray diffraction pattern of the glass samples

The structural analysis of glass samples describes by FTIR measurements. Figure 2 shows the FTIR spectra of glass composition 48 Li_2O – 48 B_2O_3 – (4-x) Cr metal –x Cu metal [x= 0, 1, 2 and 3] mol %.

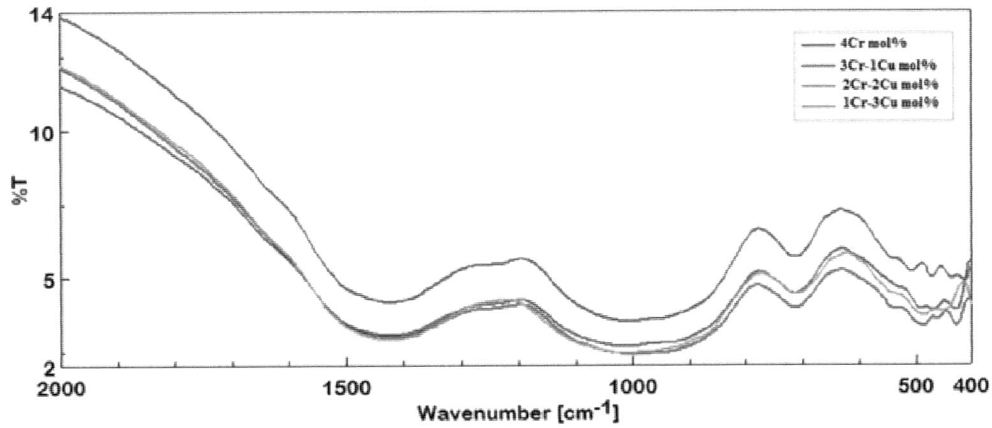

Figure 2 the FTIR spectra of glass samples

The functional groups of borate glasses observed in three IR spectral regions[4][12]. (I) The range is 1600cm^{-1} to 1200cm^{-1}, which assigns to the BO3 vibration group (containing connecting and non-bridging oxygen). ii) The range between 1200 cm^{-1} and 800 cm^{-1} due to the Boron tetrahedral vibration units. (iii) From 800 cm^{-1} to 600 cm^{-1} because of the bending vibrations of triagonal and tetrahedral of borate groups.
From figure 2, the band at 412 cm^{-1} [15][16] appears in the spectrum of glass sample containing 4 mol % Cr may be due to vibrations of $(CrO_4)^{2+}$ structural units and vibrations of Cr_2O_3 units.
This band shifted to the wave number 428 cm^{-1} with the increase of copper substitution and overlap with the band at 435 cm^{-1} that represent the vibrations of CuO_6 units[12]. The band at 485 cm^{-1} attributed to the B—O—B linkage[17]. From the FTIR result, it may understand that the transition metal may bridge with the borate band and formed M—O—B band (M is the transition metal).

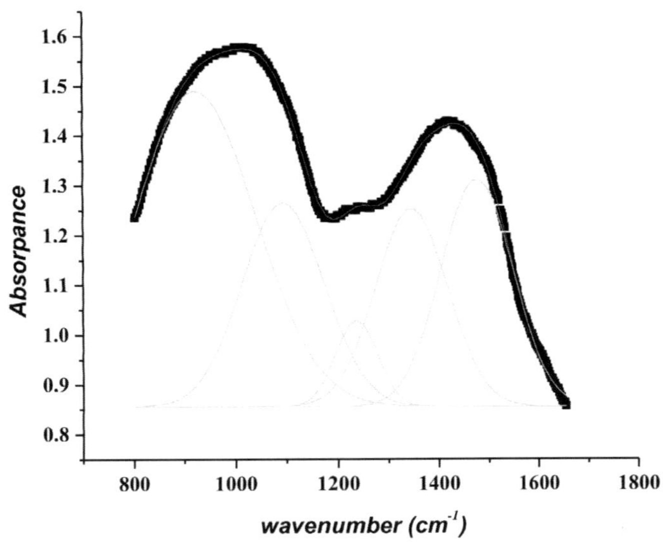

Figure 3 the deconvolution in Gaussian bands of the infrared spectrum for the glass containing 4 mol% Cr.

The broad FTIR band of the glass was attributed to the overlapping of some individual bands with each other. Gaussian deconvolution converted the broad infrared spectra into groups of sections for each individual to reflect the more information about their structural groups.
For instance, the deconvolution in Gaussian groups as shown in figure 3 for the glass containing 4 mol % copper metal. It is plainly the N_4 (the ratio between boron tetrahedral concentration to the BO3 and BO4) decreasing with the increasing the copper metal content[18]. The reduction in the number of BO_4 groups with the substitution of Cr by copper depending on the amount of oxygen and changed BO_4 to BO_3 groups and form non-bridging oxygen (NBO). Which demonstrate that the copper metal enters the glass as a modifier until 2 mol %, and after that enters as network former and a modifier.
Figure 5 shows the change in the relative area under the FTIR curves in the region containing BO_4 units have the inverse shape in the region containing BO_3 units, meaning that Cu replacement Cr and enters the borate matrix to form BO_4 units at the expense of BO_3 units without forming non-bridging oxygen.

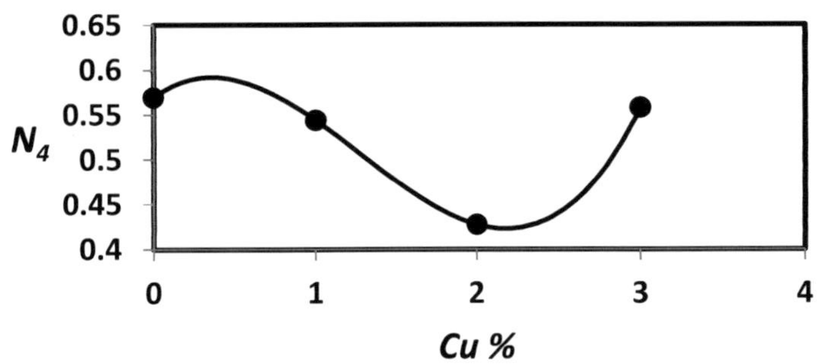

Figure 4 the N4 value as a function of composition.

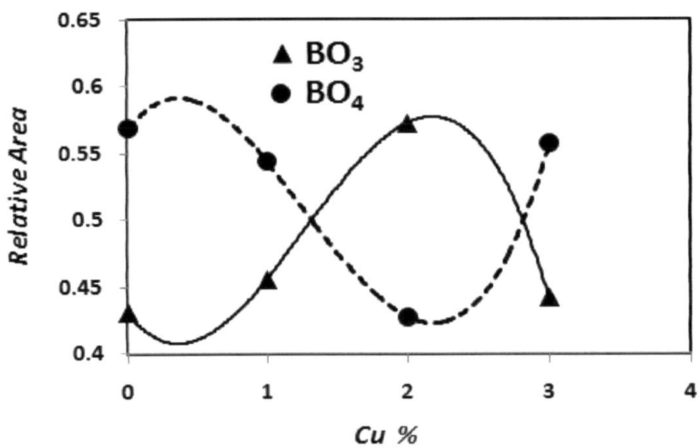

Figure 5 the change in the relative area for BO_4 and BO_3 units under the IR curves.

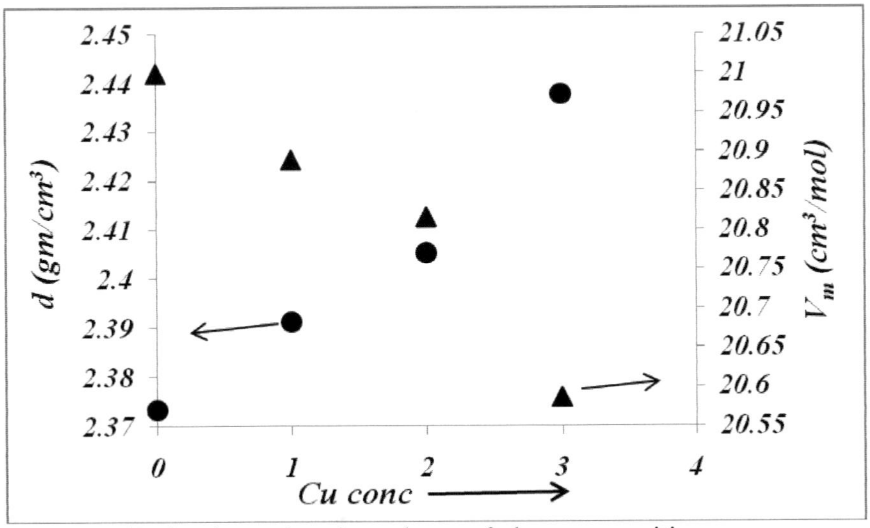

Figure 6 the density and molar volume of glass composition

The density (ρ) is a vital physical property which accustomed to investigating the tightness, changes in the structural units of material[13]. Many factors affect the density of glass samples[14]. The density increases with the substitution Cr by Cu metal as appeared in figure 6. This continued increase with the increase in copper metal content which has a higher molecular weight. From figure 6 it very well may watch that the molar volume decrease, which caused the glass to become more compact. The structure compaction is more obvious conduct of the molar volume curve. The molar volume[14] can compute from the density estimations communicated in the relation.

$$V_m = \frac{M_w}{\rho} \ \text{cm}^3/\text{mol}$$

Where M_w is the total atomic weight of the component of glass system. The molar volume decreases due to the reduction of non-bridging oxygen[19]

IOP Conf. Series: Materials Science and Engineering **956** (2020) 012013 doi:10.1088/1757-899X/956/1/012013

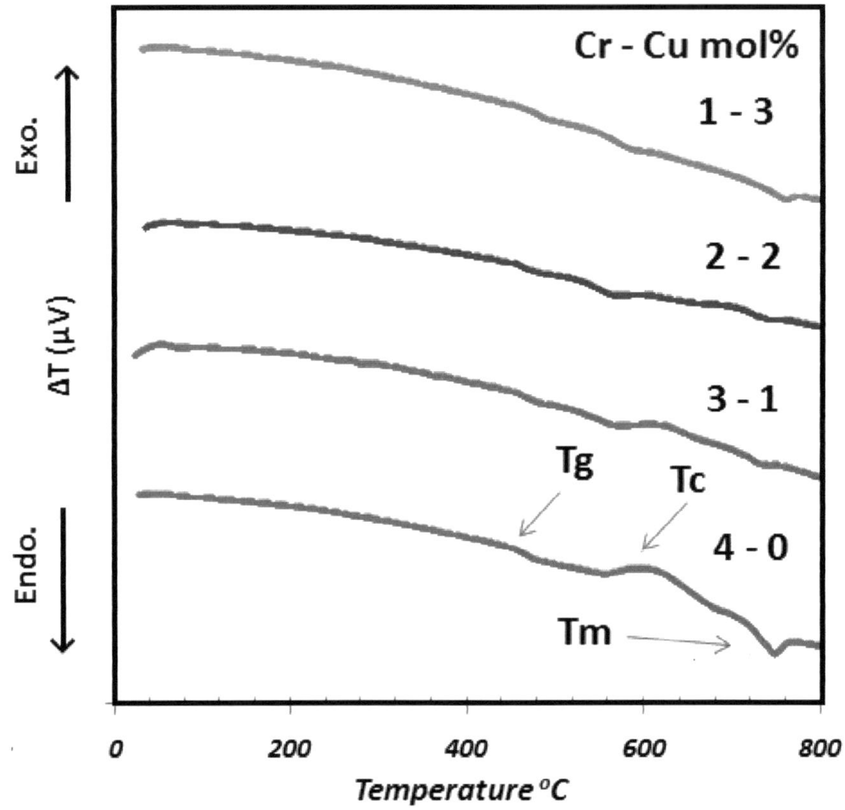

Figure 7 the DTA of lithium borate glass with the composition 48 Li_2O–48 B_2O_3–(4-x) Cr metal–xCu metal [x= 0, 1, 2 and 3] mol %

The thermal stability and glass framing ability of glass characterized by the glass transition temperature (Tg) and the crystallization temperature (Tc)[20]. Figure 7 outlines the differential thermal analysis, investigation curves of glass samples. The estimations of thermal for glass samples decided from curves and recorded in table 1. The glass transition temperature depends on numerous factors[21]. From table 1, it can see that the sample containing 3mol% Cu has a larger value of Tg that is illustrated as the sample containing more BOs than the other samples. The thermal parameters determined from the DTA result obtained, such as (S=Tc-Tg and Hrubý ($H_R = \frac{(Tc-Tg)}{(Tm-Tc)}$)[22] are useful to know the glass stability and measures the ability of a system to become glass, glass said to be thermally steady if S >100 °C. As a whole, it is difficult to prepare glasses with HR ≤ 0.1. Otherwise, the glass stability is higher with a larger HR value [22]. The obtained values of S and HR as shown in table 1, indicating that all glass samples has high glass forming ability and thermal stability.

Table 1 the DTA values and thermal parameters.

Conc(mol%)	Tg°C	Tc°C	Tm °C	S °C	Hr
4 Cr	468	605	748	137	0.96
3Cr-1 Cu	465	617	731	152	1.33
2Cr-2Cu	462	599	740	137	0.97
1Cr-3 Cu	482	612	750	130	0.94

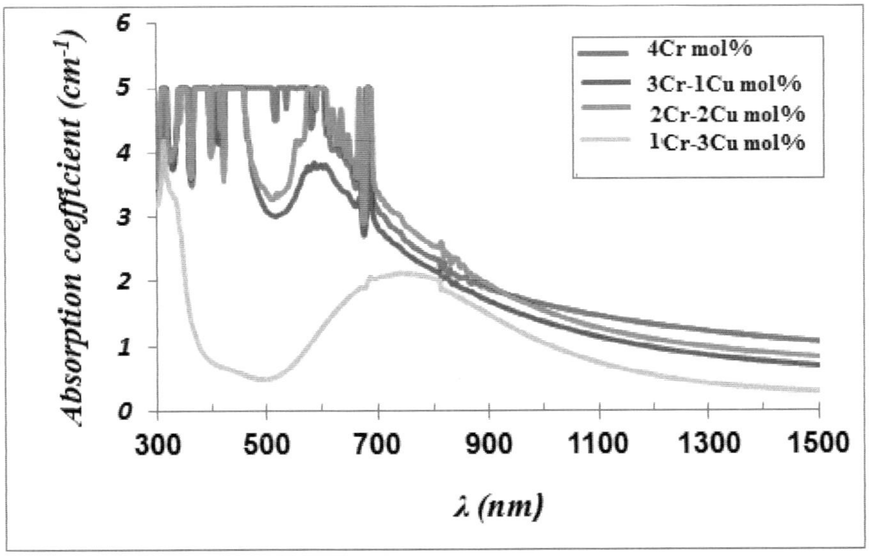

Figure 8 the optical absorption of lithium borate glass containing different concentration of copper and chrome metal.

The absorption spectra of the glasses containing 48 B_2O_3- 48 Li_2O- (4-x) Cr-x Cu where x=0, 1, 2, and 3 mol % depicted in Figure 8. From figure 8, it can see that the glass containing Cr metal has high absorption intensity and decrease with the substitution Cr with Cu metal. Broadband around 765 nm that represent the attendance of Cu^{2+} by the strong tetragonal distortion, with the transitions will give absorption bands lying together: $^2B_{1g}\rightarrow^2B_{2g}$, $^2B_{1g}\rightarrow^2A_{1g}$ and $\rightarrow^2E_g$[23].The absorption band at wavnumber less than 350 nm is due to the $Cu^{2+}\rightarrow O^{2-}$ charge-transfer band[6][24][25].The absorption spectra observed in the violet region may ascribe to $3d^{10}\rightarrow 3d^9\ 4s^1$ forbidden transitions, they may occur due to low symmetry of a local electric field around Cu^+ ions[26].The absorption spectrum exhibits three bands, two broad bands at 446 nm, 620 nm and a sharp band around 686 nm. It is characteristic for chromium ions (Cr^{3+}) in octahedral unit[10][27][28].The first broadband at 446 nm allocated to the spin-allowed transitions $^4A_{2g}$ (F)$\rightarrow^4T_{1g}$ (F) and second at 620 nm, which due to $^4A_{2g}$ (F)$\rightarrow^4T_{2g}$ (F) in the decreasing order of energy. The sharp band assigned to $^4A_{2g}$ (F) $\rightarrow^2E_g$ (G) transitions. The Cr^{6+} ion has a strong and a weak absorption bands at 358 nm and at 318 nm [25] refers to Cr^{6+}–O^- complex of $3d^0$ configurations. The characteristic band from 600 to 750 nm [27] can assign to the $^3A_2\rightarrow^3T_1$ (3F) of Cr^{4+} due to the splitting of the excited state in the lower symmetry of the crystal field. Figure 9 shows the Tanabe–Sugano diagram of tetrahedral coordinated Cr^{4+} ion. From Figure 9 we see an ideal tetrahedral position, the lowest free ion level for Cr^{4+} ion, ^3F divided into three components, $^3F\rightarrow\ ^3A_2 +\ ^3T_2 +\ ^3T_1$, with the energy of these levels being $^3A_2 <\ ^3T_2 <\ ^3T_1$[14].Also, from the previous figure it is evident that when Cr^{4+} is at the high crystalline field sites, the $^3T_2$ level is upper the 1E level, while at the low field sites the arrangement is reversed[12]. The energy transition of ($^4A_2$(f)$\rightarrow\ ^4T_2$(f)) give the ten times of crystal field splitting parameter Dq[14][29] and the Racach parameters (B, C) for lithium borate glass containing Cu-Cr calculated from the measured energies of the absorption bands according to the following relation[14][30] tabulated in table 2

$$\mathbf{Dq} = \frac{E(4_{A_2}) - E(4_{T_2})}{10}$$

$$\mathbf{Dq/B} = \frac{15(X-8)}{X^2 - 10X}$$

Where

$$\mathbf{X} = \frac{E(4_{A_{2g}} \rightarrow 2_{T_{1g}}) - E(4_{A_{2g}} \rightarrow 4_{T_{2g}})}{Dq}$$

The C-value evaluated utilizing the relation:

$$\frac{C}{B} = \frac{E(4_{A_{2g}} \rightarrow 2_{E_g})}{B} - 7.9 + 1.8\left(\frac{B}{Dq}\right)$$

Table 2 the values of Dq, Racah parameter (B and C), nephelauxetic parameters (h and β), the cutoff wavelength, optical band gap and urbach energy.

Conc(mol%)	Dq	Dq/B	Bcm^{-1}	C cm^{-1}	h	β	λc nm	Eg	ΔE
4 Cr	1666.7	3.92	424.77	3771	2.59	0.46	640	1.93	0.67
3Cr-1 Cu	1666.7	2.89	576.59	3418	1.77	0.63	690	1.797	0.6
2Cr-2Cu	1639.3	2.76	594.13	3375	1.68	0.65	440	1.67	0.6
1Cr-3 Cu	--	--	--	--	--	--	330	3.75	0.33

It can notice from the table that Dq / B decreased with increasing copper content indicating that the band became broader causing a weakening of the crystal field at the Cr^{3+} ionic site[29]. According to the Tanabe-Sugano diagram for Cr^{3+} ions as shown in figure 9, demonstrates that the Dq / B ratio influences the relative positions of the excited $^4T_{2g}$ and ^2Eg states of Cr^{3+} ions.

The Cr^{3+} can center as a high or low field sites in the luminescence and can recognize by the shape of the emission lines and the value of Dq / B. When the emission shows the narrow lines that the energy of $^4T_{2g}$ is higher than that of the ^2E_g state (Dq / B > 2.3) due to the $^2E_g \rightarrow {}^4A_{2g}$ (R1-line) and $^2T_{1g} \rightarrow {}^4A_{2g}$ (R2-line) spin-forbidden transitions. The Cr^{3+}ions in the high-field sites that produce yield conventional laser action. The energy of $^4T_{2g}$ is lower than that of the ^2E_g state (Dq / B < 2.3) when the emission shows the broadband.

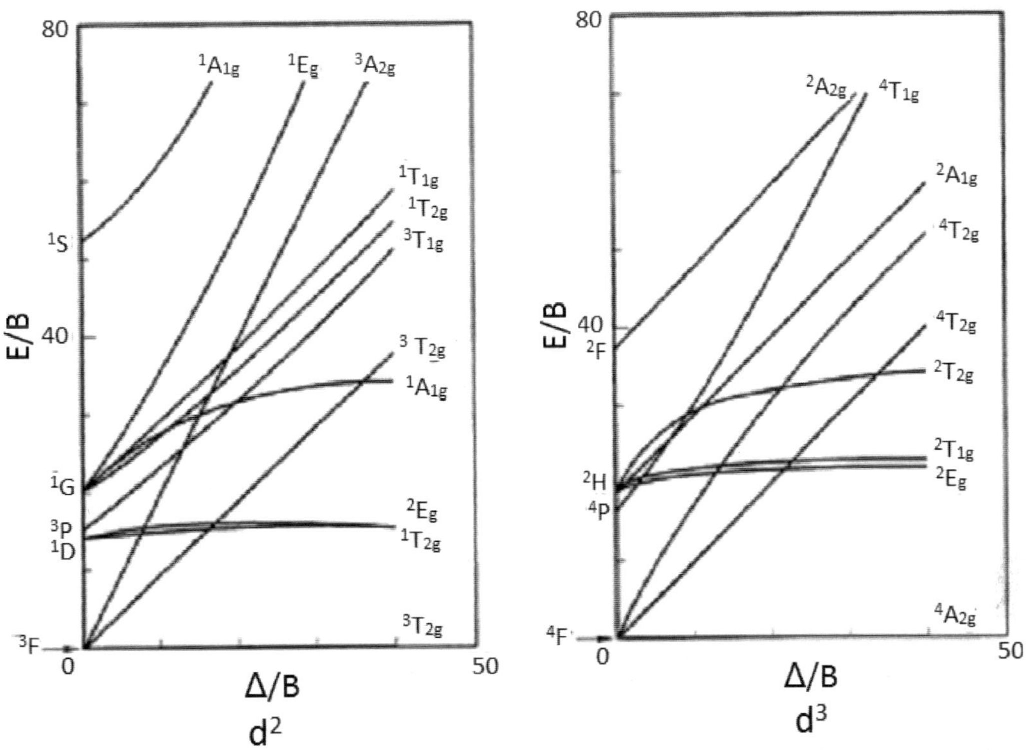

Figure 9 the Tanabe–Sugano diagram of tetrahedral coordinated Cr^{4+} ion.

The Cr^{3+}centers in glasses in the low-field sites that yield tunable laser action in disordered hosts[9][31][32]. From the values acquired from Dq / B, it can watch that the values bigger than 2.3 that show the luminescence consists of the sharp lines (R lines) that relate to parity- and spin-forbidden $^2E_g \rightarrow ^4A_{2g}$transition decaying with ms timescale. The Racah parameter B indicates the inter-electronic repulsion in the d-shell[28], from table 2 it can watch the values of B increase as the increase of Cu concentration. The larger merit of B, it means that the more electrons localized on the copper metal ion. Also, from table 2 the values of B of all glass samples lower than the B value of free ion value of Cr^{3+} (B_{free} = 918cm^{-1}).The value of B used in study the type of the Cr–O bonding (ionic or covalent bond). By using the relation[33].

$$\mathbf{h} = \left(\frac{(B_{free}-B)}{B_{free}}\right)/k_{Cr}^{3+}$$

Where h and k are nephelauxetic parameters of the ligand and the metal ion, respectively. The k For Cr^{3+} ion equal 0.21. The decrease in the h esteems represents the lessening in a covalent bond in the glass network and increment in the localization of the d-electrons due because of orbital overlapping with ligand orbital[30].

The nephelauxetic ratio, β = B (complex) /B (free ion)[34] assessed for all the glasses exhibited in Table 2 as the β values increase with the decrease of Cr metal concentration. These outcomes demonstrate a slow lessening in the covalent environment for chromium ions[34].

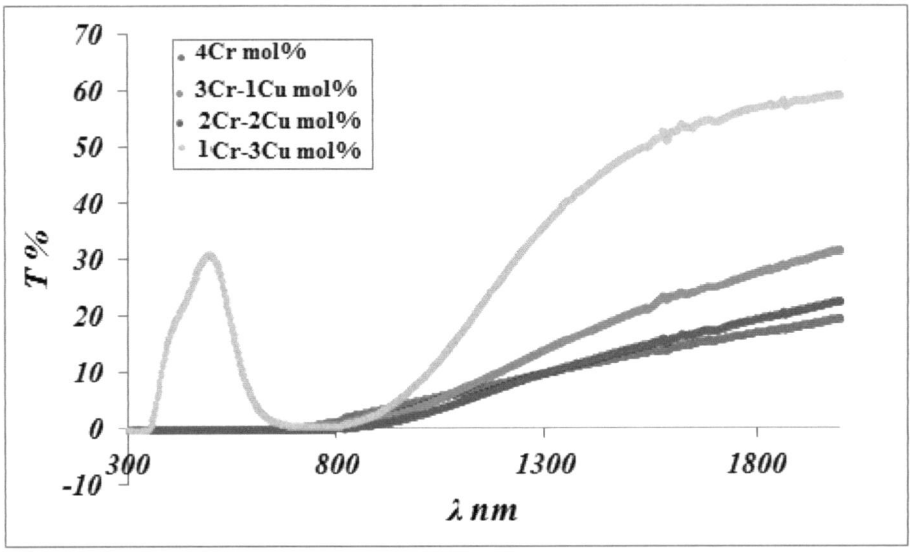

Figure 10 the transmission spectra of glass composition

Figure 10 demonstrates the optical transmission spectra for all studied borate glasses. An unmistakable cutoff (λc) has watched for all samples firmly depends on Cr-Cu substitution content and varies from 640 nm to 810 nm (Table 2). The move in λc proposes that the concentration of NBO in the glass network increment with Cu-Cr substitution[35]. From figure 10 it very well may watch that the glass sample containing 3 mol % Cu act as a narrow band color filter from 350 nm to 710 nm[12]
The estimations of the optical band gap can determine by numerous courses as from the relation[7].

$$Eg=hc/\lambda$$

Where Eg is energy gap, h is Plank's constant, c is the speed of light, and λ is the wavelength of incident photons.
The Urbach energy can calculate from the following equation[33]

$$\alpha(\nu) = A \exp (h\nu/\Delta E)$$

A is a constant, ΔE is the Urbach energy and ν is the frequency of radiation. The values of ΔE can dictate by taking the reciprocal of the slant of Ln (α) vs. λν plot.
The variation in Eg values on the variation of Cr- Cu concentration can comprehend as the structural changes occurring in the glass samples. It can observe as the glass sample containing 3 mol% of Cu have a large value of Eg, the small value of ΔE and this result agree with the literature[21]. The substitution of Cr with Cu content introduces additional defect states [4] like color centers, wrong bonds and short-range distortion in the glass structure make the Eg and ΔE values to decrease. These results agree with the values of h and β that indicate the more electrons localized around the d orbital that decreases the number of bridging oxygen around the metal.

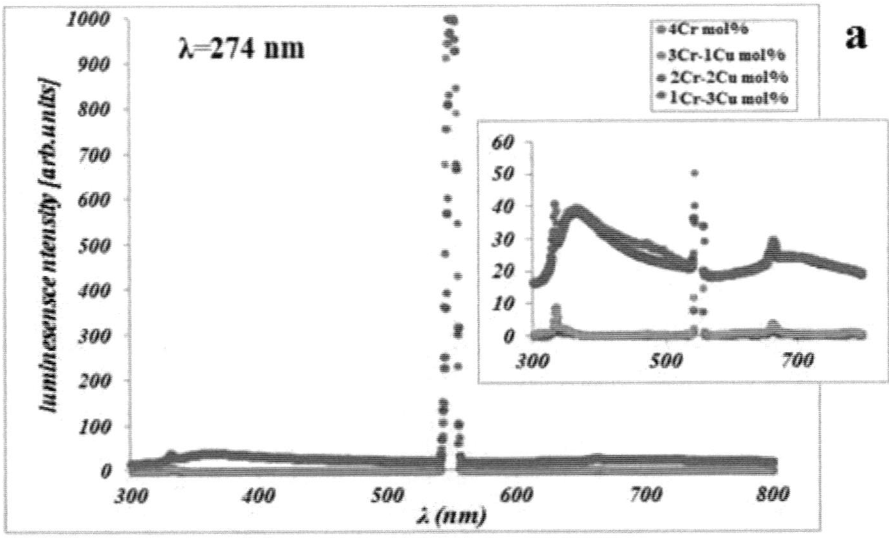

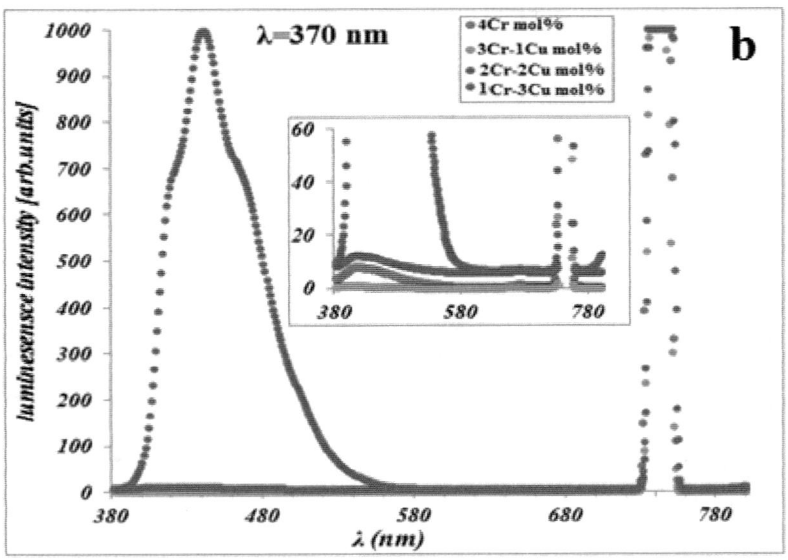

Figure 11 the photoluminescence of glass sample with excitation at a) λexc=274nm and b) λexc=370nm.

Figure 11 shows the photoluminescence of borate glass containing Cr and Cu metal with different concentration excited at (a) 274 nm and (b) 370 nm. From the figure, 11a observed that the high intensity emitted band at 540 nm due to chrome as the transitions $^4T_{1g} \rightarrow {}^4A_{2g}$ spin-allowed transitions emit green emission[14]. Figure 11b shows the high intensity emitted band at 730 nm attributed to the transitions from $^4T_{2g} \rightarrow {}^4A_{2g}$ spin-allowed transitions and $^2E_g \rightarrow {}^4A_{2g}$ spin forbidden transition (red emission) from Cr metal[14]. From figure 11 observed that the emission peaks changed from the lower energy to higher energy with the increase of excitation energy it due to these bands belong to different types of activator centers.

Moreover, the emission spectra of Lithium borate glasses containing Cr^{3+} excited at at 274 and 370 nm (36 497 cm^{-1} 4.53 eV). Showed strong visible emission in the green and orange-red region is due to d-d transitions within the Cr^{3+} formation.
The sample free from Cu and containing Cr metal excited (3.35eV) 370 nm emitted band at (740 nm 1.68 eV) attributed to the transitions from $^4T_{2g} \rightarrow {}^4A_{2g}$ spin-allowed transitions and $^2E_g \rightarrow {}^4A_{2g}$ spin forbidden transition (red emission). And observed the emission intensity has small increases as the chromium replacement by copper, which is due to the partition between the chromium ions diminishes as the quantity of Cr ions diminishes[14][36].The Cr^{3+} ion in the high field local surroundings, the emission of the Cr^{3+} ions consists of the sharp lines connected with parity- and spin-forbidden $^2E_g \rightarrow {}^4A_{2g}$ transition[31].From the emission spectra and Tanabe–Sugano diagram can prove that the emission from Cr^{3+} state. From the figure 11 observed that the blue emission band at 440 nm from the excitation at 370 nm and weak emission band at 365 nm when the samples excited at 274 nm, in addition, the intensity increase with the increase of copper concentration. The two bands belong to the emission of copper due to the $^3D_1 \rightarrow {}^1S_0$ transition of isolated Cu^+ ions and the emission from Cu^{2+}[4][36][37][38]respectively. An addition to the sample containing 3mol% Cu emitted broadband at 440 nm (2.82eV- 22727 cm^{-1}) due to the $^4A_{2g}$ (F)$\rightarrow {}^4T_{1g}$ (F) transitions. Ignatouych et al. contemplated the emission from copper ions doped lithium tetraborate glasses. They obtained a broad emission band at 460 nm. This band attributed to $3d^9\,4s \rightarrow 3d^{10}$ triplet transitions in Cu^+ ions[39] . The same results obtained by Padlyak et al.[24][40].The broadband emission at 440 nm of the sample containing 1Cr and 3Cu deconvoluted using Gaussian curves to determine the spectral position of all transitions (see the deconvolution of the band emission in Figure 12).

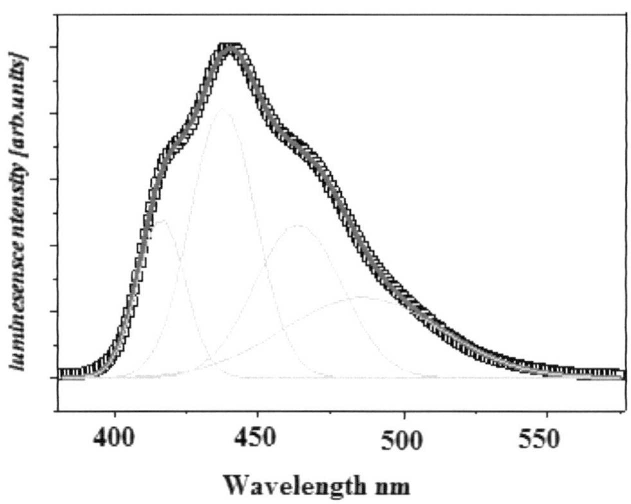

Figure 12 the deconvolution in Gaussian bands of the emission spectrum for the glass sample containing 1Cr and 3Cu mol % metal.

From Figure 12, it tends to see that the broadband emission deconvoluted to four bands at 417, 438, 464 and 486 nm different shapes (broad or sharp) and in the intensity (less or more) indicating that the more transition dominant. The emissions at 417 and 486 nm are due to Cu^{2+} ions in the glass. Be that as it may, as indicated by

IOP Conf. Series: Materials Science and Engineering **956** (2020) 012013 doi:10.1088/1757-899X/956/1/012013

the before investigations, emission at 493nm emerges in view of a confined excitation of an isolated Cu^{2+} ion[41]. The broad emission is due to, or resulting from, the transition between degraded T_{1g} / T_{2g} levels to the ^1Ag level of copper ions Cu^+, which arises from the interaction of non-bridged oxygen atoms bound by P with Cu^{2+} ions in the glass[42][43].Subsequently, in the luminescence studies, the presence of Cu^+ ions in the glass has inferred. With the addition of copper the center of emitted not affected. And then the intensity may be small increases.

4. Conclusion

Lithium borate glass samples containing copper and chromium metal synthesized by the melt quenching technique. XRD investigate the amorphous nature of glass samples. The FTIR investigation demonstrates that the Cu substitution Cr and enters the network structure and form BO_4 units at the expense of BO_3 units without forming non-bridging oxygen. Differential thermal analysis parameter of glass samples that demonstrates all the samples have the glass forming ability and thermal stability. The optical properties of glass samples containing copper and chromium metal obtained. The values of Dq / B represents that Cr^{3+} ions in a high field and decrease with the copper content increase. Furthermore, the green and red emission due to the transitions from $^4T_{1g}$ and $^4T_{2g}$ excited states to the ground state of Cr^{3+} ions. By the addition of Cu up to 3 mol%, the highest intensity variation observed at 440 nm (blue) band for transition in glass sample containing 1Cr-3Cu
As a result, interesting changes are observed in the emission spectra which correlate with changes in the glass lattice structure.

IOP Conf. Series: Materials Science and Engineering **956** (2020) 012013 doi:10.1088/1757-899X/956/1/012013

References

[1] Nalin M, Poulain M, Ribeiro S. J. L., and Messaddeq Y 2001 Journal of non-crystalline solids**284** 110–116
[2] Soga N, Hirao K, Yoshimoto M, and Yamamoto H 1988 Journal of applied physics**63** 4451–4454
[3] Balachander L, Ramadevudu G, Shareefuddin M, Sayanna R, and Venudhar Y. C 2013 *Science Asia* **39** 278–283
[4] Bhogi A, Kumar R. V, and Kistaiah P 2015 Journal of non-crystalline solids**426** 47–54
[5] Marzouk M. A, ElBatal F. H, and Abdelghany A. M 2013 SpectrochimicaActa Part A: Molecular and Biomolecular Spectroscopy**114** 658–667
[6] Stefan R, Culea E, and Pascuta P 2012 Journal of non-crystalline solids**358** 839–846
[7] Obayes H. K, Wagiran H, Hussin R, and Saeed M. A 2016 Materials& Design**94** 121–131
[8] Krishna G. M, Gandhi Y, Venkatramaiah N, Venkatesan R, and Veeraiah N 2008 Physica B: Condensed Matter**403** 702–710
[9] Lesniewski T ,PadlyakB V, JustynaMahlikBarzowska,SebastianAdamivV T, Nurgul Z, GrinbergMarek, 2016 Optical Materials *(Amst).***59** 120–125
[10] Kaczmarek S. M 2002 Optical Materials *(Amst).***19** 189–194
[11] Ikeda H, Murata T, and Fujino S 2015 *Mater. Chem. Phys.***162** 431–435
[12] Kashif I and Ratep A 2015 Journal of Molecular Structure**1102** 1–5
[13] Shailajha S, Geetha K, VasantharaniP, and KadharS. P. S. A 2015 SpectrochimicaActa Part A: Molecular and Biomolecular Spectroscopy**138** 846–856
[14] KashifI and RatepA 2016 Optical and Quantum Electronics **48** 516
[15] CozarO, ArdeleanI, BratuI, IloncaG, and Simon S 1993 *Solid State Communcation*, **86** 569–572
[16] YusubS and RaoD. K 2014 Journal of non-crystalline solids**398** 1–9
[17] SumalathaB, OmkaramI, RaoT. R, and RajuC. L 2011 Journal of non-crystalline solids**357** 3143–3152
[18] KashifI, Abd El-ghanyA, Abd El-MaboudA, ElsherbinyM. A, and SanadA. M 2010 Journal of Alloys and Compounds.**503** 384–388
[19] KumarD, RaoS. M, and SinghS. P 2017 Journal of non-crystalline solids**464** 51–55
[20] DalalS *et al.* 2015 Materials Research Bulletin**70** 559–566
[21] KashifI and RatepA 2015 *Appllied Physics A***120** 1427–1434
[22] MharebM. H. A *et al.*2016 Journal of Luminescence**177** 366–372
[23] MoustafaY. M, HassanA. K, El-DamrawiG, and YevtushenkoN. G 1996 Journal of non-crystalline solids**194** 34–40
[24] PadlyakB *et al.* 2010 Journal of non-crystalline solids**356** 2033–2037
[25] A. ElBatalH, AbdelghanyA. M, and AliI. S 2012 Journal of non-crystalline solids**358** 820–825
[26] RaoL. S, ReddyM. S, RaoD. K, and VeeraiahN 2009 *Solid state Science***11** 578–587
[27] MaH, LiuP, YangQ, DengD, and XuS 2014 *Materials Letter***116** 13–15
[28] KesavuluC. R, ChakradharR. P. S, MuralidharaR. S, RaoJ. L, and AnavekarR. V 2010 Journal of Alloys and Compounds**496** 75–80

[29] ChakradharR. P. S, MuraliA, and RaoJ. L 1998 Journal of Alloys and Compounds **281** 99–107
[30] AhmadF 2014 Journal of Alloys and Compounds**586** 605–610
[31] PisarskiW. A, PisarskaJ, Dominiak-DzikG, and Ryba-RomanowskiW 2009 Journal of Alloys and Compounds**484** 45–49
[32] PadlyakB. V, Ryba-RomanowskiW, LisieckiR, AdamivV. T, V BurakY, and TeslyukI. M 2012 *Optical Materials (Amst).***34** 2112–2119
[33] Sundari G. R *et al.* 2014 *Optical Materials (Amst).***36** 1329–1335
[34] Naga RajuG, VeeraiahN, NagarjunaG, and SatyanarayanaP. V. V 2006 Physica B: Condensed Matter**373** 297–305
[35] RaniS, SanghiS, AhlawatN, and AgarwalA, 2014 Journal of Alloys and Compounds**597**, 110–118
[36] KaurS, KaurP, SinghG. P, KumarS, and SinghD. P 2015 *Optical Materials (Amst).***47** 276–284
[37] V NikonorovN, SidorovA. I, TsekhomskiiV. A, and ShakhverdovT. A 2013 *Opt. Spectrosc.***114** 379–383
[38] BabkinaA. N, V NikonorovN, SidorovA. I, ShirshnevP. S, and ShakhverdovT. A 2014 Optics and spectroscopy**116** 84–90
[39] PedriniC 1978 *Physica status solidi***87** 273–286
[40] SrinivasuluK, OmkaramI, ObeidH, KumarA. S, and RaoJ. L 2012 *Physica B Condensed Matter***407** 4741–4748
[41] ThulasiramuduA and BuddhuduS 2006 Journal of Quantitative Spectroscopy and Radiative Transfer**97** 181–194
[42] RamkumarJ *et al.* 2008 Journal of non-crystalline solids**354** 1591–1597
[43] ChanshettiU. B, SudarsanV, JogadM. S, and ChondhekarT. K 2011 *Physica B Condensed Matter***406** 2904–2907

IOP Conf. Series: Materials Science and Engineering **956** (2020) 012014 doi:10.1088/1757-899X/956/1/012014

Gold Nanorods as Contrast Agent for Photoacoustic Imaging (PAI) of Breast Cancer

H Awad[1], T Abdallah[1], K Easawi[2], S Negm[2], and H Talaat[1*]

[1] Physics Department, Faculty of Science, Ain Shams University, Abbassia, Cairo, Egypt
[2] Department of Mathematical and Physical Engineering, Faculty of Engineering (Shoubra), Benha University, Cairo, Egypt

* E-mail: hassantalaat@hotmail.com.

Abstract. Gold Nanorods (AuNRs) can play a fundamental role in the enhancement of photoacoustic imaging (PAI). This is due to their high and tunable optical absorption cross-section. Normally, the most convenient and highly stable AuNRs are synthesized using Cetyltrimethylammonium bromide (CTAB) as a surfactant. In this work, surface modification with poly ethylene glycol (PEGylation) and Silica coating (SiO_2-coating) were applied to help in reducing toxicity of CTAB capped AuNRs. Breast cancer cell line (MDA-MB-231) was used to assess the modified AuNRs as contrast agents. A much brighter PAI of cells loaded with SiO_2-coated AuNRs was obtained. Such an enhanced PA image is due to the larger cellular uptake, as observed from bright field microscope images taken for the cells. On the other hand, the corresponding PA image for cells with PEGylated AuNRs is diminished. The increase in cell uptake, in case of SiO_2-coated AuNRs, could result due to the adsorption of protein, contained in the cell culture media, on the negatively charged silica surface, which increases intercellular uptake by receptor-mediated endocytosis. Moreover, the increase of the brightness of PAI of SiO_2-coated AuNRs may be due to the lowering of the interfacial thermal resistance between gold and the surrounding solvent. This allows the particle to release more heat to its environment which increases the generated acoustic signals from these particles.

Keywords: Photoacoustic Imaging, contrast agents, Gold nanorods, Cancerous cells.

1. Introduction

Photoacoustic imaging (PAI) is a biomedical imaging technique based on the photoacoustic effect discovered by A.G. Bell in 1882(1). The photoacoustic effect refers to the conversion of light energy into acoustic energy. In PAI, a non-ionizing laser pulses are delivered into the biological tissue, some of the delivered energy will be absorbed and then converted into heat energy. This leads to transient thermoelastic expansion and thus wide band ultrasonic emission. The generated waves are detected by ultrasound transducer and then analyzed to produce images. The motivation for PAI is to combine ultrasonic resolution with high contrast due to light absorption. Thus, in the PAI ultrasound is detected and not light. The ultrasound energy experience less scattering and attenuation, therefore, a good spatial distribution of optical absorption can be obtained in deep tissues with high resolution and high penetration depth compared to other optical imaging modalities. Photoacoustic imaging and microscopic applications shows a great potential in ex-vivo, in-vivo, and in-vitro settings, due to its advantageous technique compared to other imaging techniques. The optical absorption is the primary contrast mechanism in photoacoustics, as it possesses information regarding the presence of tissue components such as hemoglobin, and melanin. Those components are considered as endogenous contrast agents. Strong optical absorption by hemoglobin and oxy-hemoglobin allows the visualization of blood vessels. Furthermore, those contrast agents have a specific absorption spectrum which makes them distinguishable from other tissue components. Their presence gives more pathological and functional information about the imaged tissue. On the other hand, many exogenous contrast agents were used to enhance the photoacoustic signal. Gold nanoparticles are very interesting PA contrast agents (2-4). This is because they have a large absorption cross-section, due to the surface plasmon resonance (SPR) effect. Moreover, they have a high tunability as the absorption band position can be controlled by changing the nanoparticle size and shape. AuNRs are more preferred, because they have strong longitudinal SPR band lying in the optical biological window (700-1000nm), where the absorption of light by tissue is minimal. This condition is important for PAI of deep tissues that cannot be obtained by many other imaging methods Therefore, AuNRs (5) have been developed and applied in both in-vitro and in-vivo applications. They are also functionalized for further enhancement in PAI contrast (6-10). AuNRs aspect ratios (length/width) between 3 and 4 are favored for PAI because their longitudinal SPR band lies at the biological window (11-19).

Silica (SiO_2) coating is considered as one of the most chosen functionalization methods for AuNRs. It is applied in many applications like PAI, cell-tracking, targeted drug delivery, and Surface Enhanced Raman Spectroscopy. This is due to their high stability and biocompatibility and functionality, compared to conventional CTAB capped AuNRs (20-22). In case of SiO_2-coated rods, the optical properties does not change in such techniques using a highly powerful laser, while in the case of CTAB capped AuNRs, melting takes place and there will be a shift in the longitudinal surface plasmon resonance band towards visible (23). Moreover, SiO_2-coating prevents particle aggregation and plasmon coupling effect at high particles concentration. Also, the surface area of particles increases due to silica coating, this help in more conjugation of antibodies or any other moieties.

In this work, three gold nanorods (AuNRs) samples were prepared via the seed mediated growth technique(12), where silver nitrate ($AgNO_3$) with different amounts were added to change the rods size. The PA image for particles in gelatin phantom is taken at different wavelengths using optical parametric oscillator (OPO) Laser, to obtain the wavelength for the maximum acoustic signal, for each particle size in gelatin inclusions. PA images are taken for MDA-MB 231 breast cancer cells loaded with PEGylated AuNRs and another cell loaded with SiO_2-coated AuNRs to assess the enhancement of PA image in each case.

2. Experiment

2.1. Synthesis and characterization of AuNRs with different sizes

The AuNRs were prepared using the seed-mediated growth technique (12). The seed solution was obtained by mixing 5mL of 0.2 M (CTAB, Sigma) with 5 mL of 0.5 mM HAuCl4, and 0.6 mL of ice-cold 10 mM sodium borohydride (NaBH4, Sigma), was then added. This results in the formation of a brownish yellow solution. Vigorous stirring of the seed solution was continued for 2 minutes. Afterword's, the seeds were left for two hours before adding to the growth solution. The growth solution was obtained by mixing (900 mL, 0.1 M) CTAB with (50 mL, 10 mM) HAuCl4, and (15-20-25) ml of 4 mM AgNO3(Sigma) then (10 mL, 0.1M) ascorbic acid was added drop wise. The solution then becomes colorless. Finally, 10 mL of the seed solution was then added to the growth solution. The color of the solution gradually changed within 20 minutes. The solution is set for overnight before concentration. We thus obtained three AuNRs solution with different longitudinal plasmon resonance peaks. The AuNRs were centrifuged twice at 11,000 rpm for 10 minutes and then re-dispersed in deionized water to remove the excess CTAB molecules.

CTAB, on the surface of the AuNRs was replaced by methoxy thiolated polyethylene glycol (mPEG-SH, MW 5000, Laysab Bio.) through ligand exchange. The CTAB-stabilized AuNR dispersion was added to an equal volume of (0.2 mM) mPEG-thiol aqueous solution. The solution was shaken at lowest speed overnight allowing CTAB ligands to exchange with mPEG-SH. The excess mPEG-SH was removed via centrifugation at 10,000 rpm for 10 minutes for two times. And the PEGylated AuNRs were re-suspended in water.

SiO_2-coating is carried out using the method in (21). In brief, Tetraethyl orthosilicate (TEOS) is used to get a mesoporous siloxane polymer on the surface of AuNRs. About 1.2 mL of PEGylated AuNR colloidal solution was mixed with 1.8 mL of isopropanol (IPA) under vigorous stirring, then an ammonia solution in IPA (3.84 v% ammonia (33 wt%)) was added slowly under vigorous stirring until the solution reached PH=11. Finally, (0.40

mL, 100 mM) solution of TEOS in IPA was added under gentle stirring. The reaction mixture was allowed to react for 2 hours.

The samples were characterized by UV–Vis spectroscopy (Ocean Optics), to determine their optical absorption spectra. Zetapotential (Zeta sizer Nano Series, Malvern) was measured to give an indication to the surface charge of the different particles. The particle size was determined using transmission electron microscopy (TEM), The samples were inspected using a JEOL 100 CX-II TEM operating at 100 kV. The TEM imaging of gold nanoparticles were carried out by placing the nanoparticles on carbon-coated copper grid and dried overnight in air. All of the micrographs were recorded on a CCD camera, and the images were analyzed using Image J software to get the particle size and the standard deviation.

2.2. Cell culture

MDA-MB-231 Breast cancer cell line is used to study the effect of PEGylation and Silica coating of gold nanorods as contrast agent for PA imaging. The used cell culture media consists of Dulbecco's modified eagle medium: Nutrient Mixture (DMEM/F12), 10% fetal bovine serum, and 2% pencilline/streptomycin sulphate. A 37°C with 5% CO_2 incubator was used for cell incubation. Cell media were changed every three days. 1mL of trypsin is then added to remove the cells when about 90% confluent is reached. Trypsin is an enzyme that breaks down proteins which enables the cells to adhere to the cell culture plate. Cells are then counted using a hemocytometer and reseeded in a 6-well cell culture plates, having about $5x10^5$ cell in each well, and then let to adhere overnight before nanoparticles addition.
To incubate the particles with the MDA-MB-231 cancer cells, concentrated PEGylated and SiO_2-coated rods are resuspended into phenol red-free DMEM/F12 cell media to known optical density (OD), the standard DMEM/F12 solution was aspirated from the cultured cells, and replaced with equal volume of nanoparticles in phenol red free DMEM/F12, such that the OD of the solution is 1. The cells are placed in the incubator for 24 hours at 37 °C. After incubation, excess nanoparticles were removed by gently aspirating and dispensing phosphate-buffered saline (PBS) for 3 times. Cells are then passaged with trypsin and the trypsin is quenched with cell media. The cells are then centrifuged and fixed with 10% formalin, centrifuged, and resuspended in cold PBS. A Tissue mimicking phantom with loaded cell inclusions was prepared for photoacoustic imaging. The cell suspension (20 μL) was mixed with gelatin solution (6% by weight after mixing) and placed on the base layer of the gelatin phantom.

2.3. Combined US-PA imaging

Combined US-PA images for the gelatin tissue-mimicking phantoms were obtained using Vevo 2100 ultrasound micro imaging system (VisualSonics, Inc.) described in (24). PA images for the different three AuNRs inclusions are take over the whole range of the laser source (680-930 nm). The PA spectra as a function of wavelength is obtained from the corresponding PA images by subtracting the average PA signal, calculated over the total area of AuNRs inclusion, from the background signal having the same area. PA images were taken for two gelatin cell inclusions, one having cells loaded with PEGylated AuNRs and the other one contains cells loaded with SiO_2-coated AuNRs.

3. Results and Discussion

The UV-Vis absorption spectra for the three prepared samples with different amounts of $AgNO_3$ (15ml, 20ml, 25 ml) as mentioned in the experimental part, are shown in **Figure 1**(a). The obtained longitudinal SPR peaks were at 670 nm, 767 nm, and 838 nm, respectively, while the transverse mode is at 520 nm for all samples. Figure 1(b) shows the UV-Vis spectra of PEGylated and SiO_2-coated AuNRs of sample (2) in Figure 1(a). There is about 20 nm red shift to the longitudinal surface plasmon resonance peak between the PEGylated and SiO_2-coated AuNRs, which confirms the coating process.

IOP Conf. Series: Materials Science and Engineering **956** (2020) 012014 doi:10.1088/1757-899X/956/1/012014

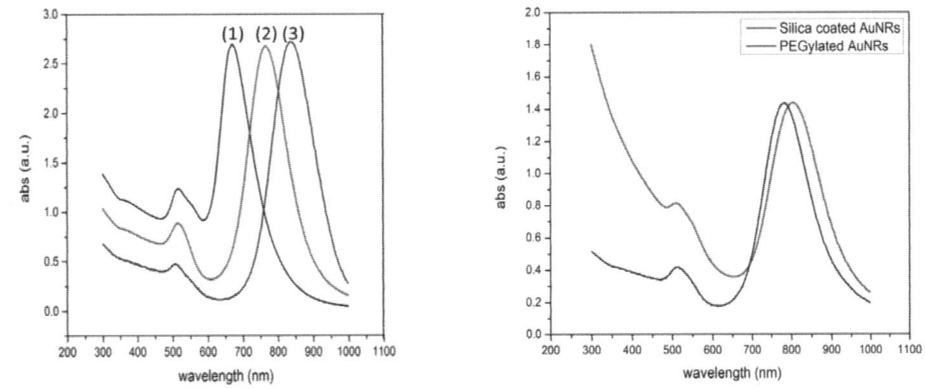

(a) (b)

Figure 1 (a) UV-Vis spectra for AuNRs with different added amounts of AgNO3 ((1):15ml, (2):20ml, and (3):25ml), (b) UV-Vis absorption of blue: PEGylated and red: silica coated gold nanorods

Figure 2(a), (b) shows the TEM images taken for the samples (1) and (2) in Figure 1(a), respectively. Figure 2(a) shows monodispersed nanorods with length = 35.93 ±5.2 nm, width =8.94 ±1.08 nm , and aspect ratio of about 3.72, Figure 2(b) gives monodispersed nanorods with length =56.71 ±5.76 nm, width =15.25 ±2.32 nm and aspect ratio of 4.02. The TEM image of the later sample coated with SiO2 of 40 nm thickness is shown in Figure 2(c).

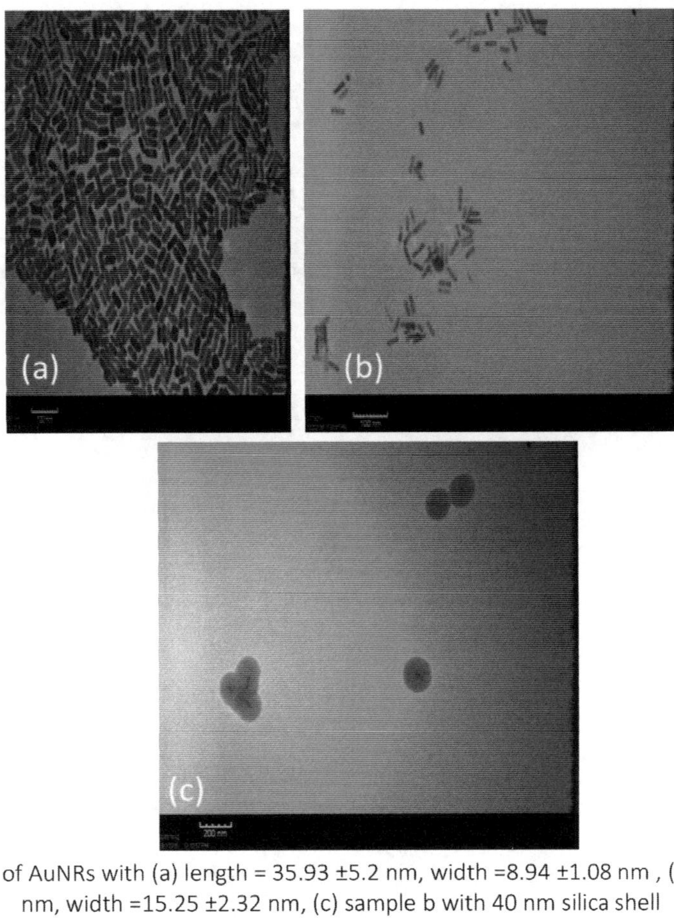

Figure 2 TEM images of AuNRs with (a) length = 35.93 ±5.2 nm, width =8.94 ±1.08 nm , (b) length =56.71 ±5.76 nm, width =15.25 ±2.32 nm, (c) sample b with 40 nm silica shell

(a) Zeta potential measurements were carried out for both PEGylated and SiO_2-coated AuNRs to confirm the coating process. (b)

Figure 3 shows that there is a change in the Zeta potential from +5.46 mV, for PEGylated AuNRs, to -45.8 mV due to the silica coating of AuNRs. The high negativity of SiO_2-coated AuNRs results in an increase of its surface free energy. This leads to better interaction of nanoparticles with proteins to reduce its surface free energy.

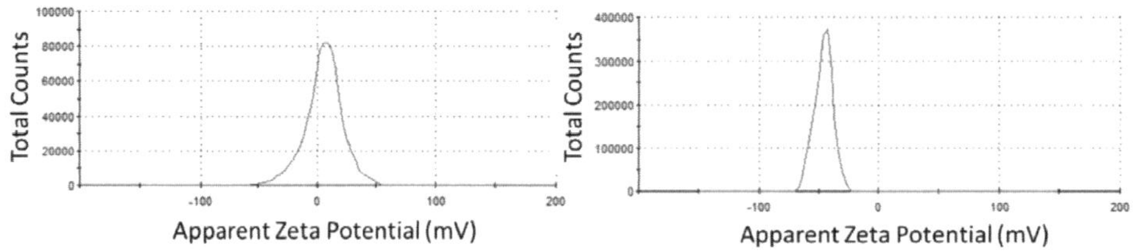

(b) (b)

Figure 3 Zeta potential of (a) PEGylated and (b) Silica coated rods respectively

Figure 4 shows the PA spectra for the prepared AuNRs. Gelatin inclusions for the three prepared AuNRs samples having different aspect ratios are prepared at the same OD (3). The photoacoustic spectra from the tissue-mimicking phantom were obtained using the combined US-PA imaging system with OPO as light source. Figure 4 also shows the corresponding US-PA overlay image for the AuNRs with three different sizes, using the laser wavelength that gives maximum acoustic signal in each case. It is clear that the PA signal is maximum at LSPR wavelength obtained by UV-Vis spectra (Figure 1).

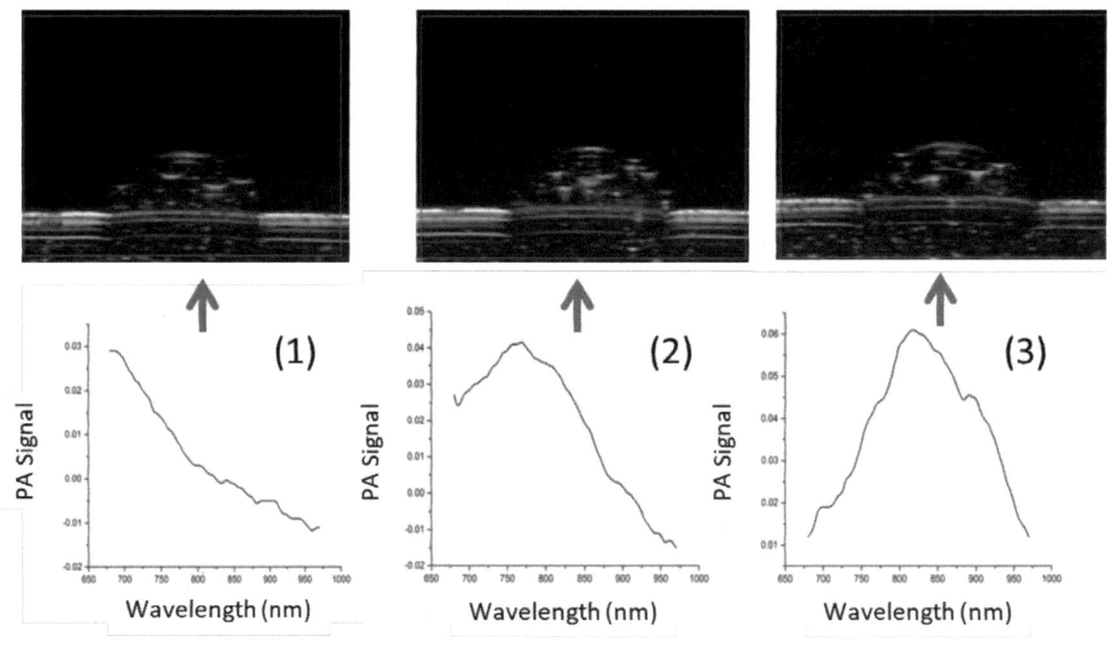

Figure 4 PA spectra for gelatin inclusions with AuNRs, with different aspect ratios shown in Figure 1a (1,2,3). Overlay US-PA images for gelatin inclusions taken at the shown corresponding peak wavelength are on top

Figure 5 shows the PA spectra for gelatin inclusions of MDA-MB-231 cells loaded with (a) PEGylated, and (b) SiO_2-coated AuNRs, as a function of wavelength (680 nm – 930 nm). It is found that the PA signal is maximized at about 805 nm, in case of cell loaded with SiO_2-coated AuNRs, while in case of PEGylated nanorods there is no signal over the whole range. An overlay US-PA images for cells loaded with PEGylated (a) and SiO_2-coated (b) AuNRs are shown in Figure 6. These images were taken at laser excitation which gives maximum PA signal.

The diminished PA signal for cells loaded with PEGylated AuNRs is due to the reduced cellular uptake of PEGylated AuNRs compared to SiO_2-coated AuNRs, as indicated from the bright field microscope images.

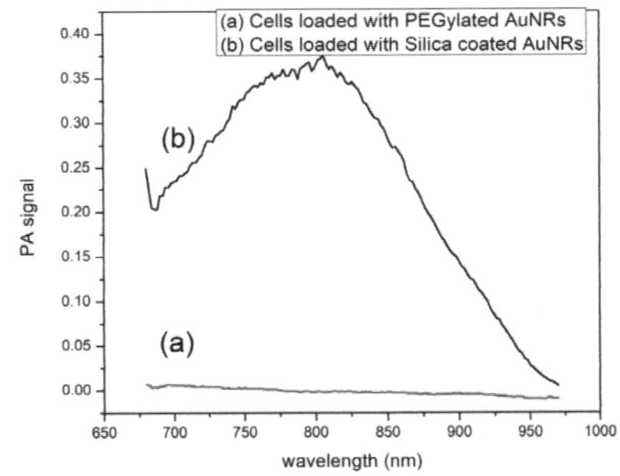

Figure 5 PA signal at different applied laser wavelengths for cells loaded with (a) PEGylated, and (b) silica coated gold nanorods

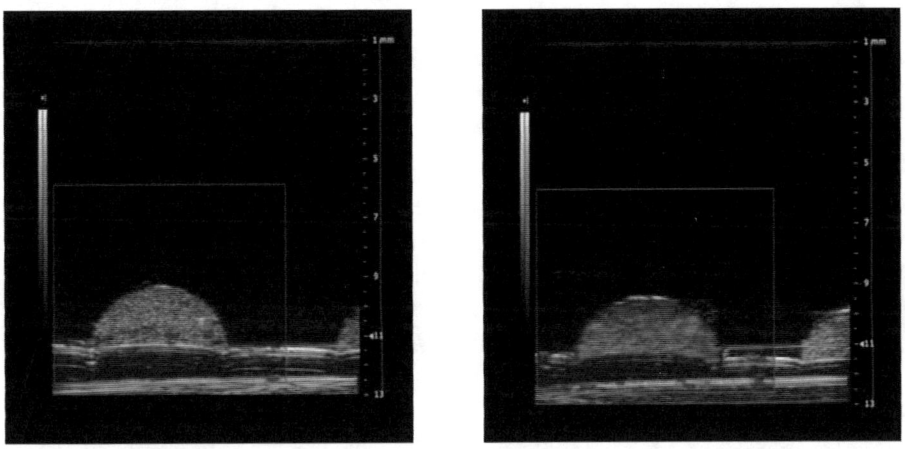

(a) (b)

Figure 6 Overlay US-PA images for cells loaded with (a)PEGylated and (b) Silica coated AuNRs

Figure 7 shows the bright field images obtained using 20x objective (0.5 NA) and Lecia 6000 DM microscope for the loaded cells. The images confirm that the cell uptake for SiO_2-coated AuNRs is more than that of PEGylated rods. It is evident that the silica coating enhances the cell uptake as well as the PA signal. This result agrees with the study of Xiao-Ming Zhu (25).

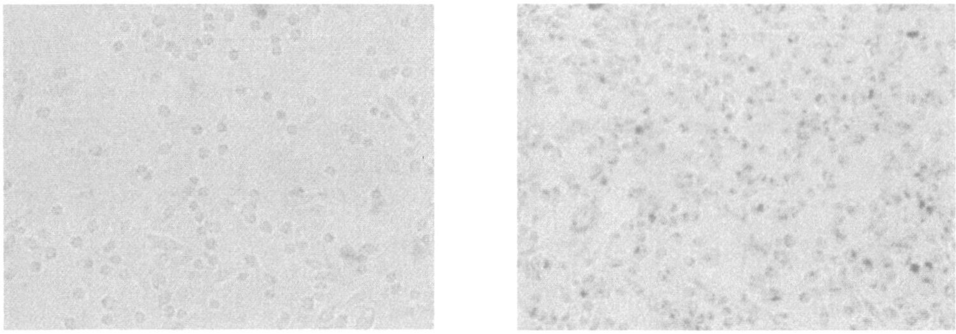

(a) (b)

Figure 7 Bright field microscope images of (a) PEGylated, and (b) Silica coated nanorods

The enhanced cellular uptake in case of SiO_2 coated AuNRs, is due to the net negative charge on the nanoparticle surface (as found in the Zeta potential measurements). This negative charge produces electrostatic interaction with positively charged function groups in proteins contained in the cell culture media via fetal bovine serum. Therefore, the protein corona that surrounds the SiO_2-coated AuNRs, induce the rods to enter the cell via receptor-mediated endocytosis. Wang L M et al (26) found that bovine serum albumin, in cell media, adsorb on surface of gold nanoparticles by about twelve Au-S bonds. On the other hand, protein complexes cannot be formed in case of PEGylated gold nanoparticles. Since the Zeta potential is almost neutral, therefore, they have a high capacity to resist the adsorption of proteins after modification with high densities of hydrophilic PEG (27) and hence, circumvent cellular uptake.

Further enhancement in the PA signal, for cells loaded with SiO_2-coated AuNRs, can be attributed to the assistance of silica shell in reduction of the interfacial thermal impedance between gold and surrounding solvent which allows particles to release more heat, leading to larger acoustic signal (28).

4. Conclusions

Though PEGylation and Silica coating help in reducing toxicity of CTAB capped AuNRs, however PEGylation diminish the PAI of AuNRs loaded cancer cells, due to reduction of cellular uptake of PEGylated AuNRs. As a result of incubating SiO_2-coated AuNRs with cancer cells, an enhanced cellular uptake is observed for silica coated AuNRs compared to the diminished PEGylated NRs. this enhanced PA image is due to the formation of a protein corona around SiO_2-coated AuNRs which increase intercellular uptake by receptor-mediated endocytosis. Moreover, the increase of the SiO_2-coated AuNRs is also because the silica coating helps lowering the interfacial thermal resistance between gold and the surrounding solvent which allows the particle to release more heat to its environment. This increases the generated acoustic signals from these particles.

Acknowledgment

The authors greatly appreciate Prof. Stanislav Emelianov, Georgia Institute of Technology for hosting H. A. at his laboratory. The authors also thank the generous support of the Egyptian STDF grant ID 3746.

References

[1] MIMS III FM 1980 *Optics News* **6** 8-16
[2] Kim G, Huang S-W, Day KC, O'Donnell M, Agayan RR, Day MA, et al 2007 *Journal of biomedical optics* **12** 044020
[3] Song KH, Stein EW, Margenthaler JA, Wang LV 2008 *Journal of biomedical optics* **13** 054033
[4] De La Zerda A, Zavaleta C, Keren S, Vaithilingam S, Bodapati S, Liu Z, et al. 2008 *Nature nanotechnology* **3** 557-62
[5] Ye X, Gao Y, Chen J, Reifsnyder DC, Zheng C, Murray CB 2013 *Nano letters* **13** 2163-71
[6] Oldenburg S, Averitt R, Westcott S, Halas N 1998 *Chemical Physics Letters* **288** 243-7
[7] Ah CS, Yun YJ, Park HJ, Kim W-J, Ha DH, Yun WS 2005 *Chemistry of Materials* **17** 5558-61

[8] Wang Y, Liu Y, Luehmann H, Xia X, Brown P, Jarreau C, et al 2012 *Acs Nano* **6** 5880-8
[9] Nehl CL, Liao H, Hafner JH 2006 *Nano letters* **6** 683-8
[10] Wang S, Huang P, Nie L, Xing R, Liu D, Wang Z, et al 2013 *Advanced materials* **25** 3055-61
[11] Xia Y, Li W, Cobley CM, Chen J, Xia X, Zhang Q, et al 2011 *Accounts of chemical research* **44** 914-24
[12] Nikoobakht B, El-Sayed MA 2003 *Chemistry of Materials* **15** 1957-62
[13] Zhu J, Yong K-T, Roy I, Hu R, Ding H, Zhao L, et al 2010 *Nanotechnology* **21** 285106
[14] Huang X, Neretina S, El-Sayed MA 2009 *Advanced materials* **21** 4880-910
[15] Link S, El-Sayed MA 1999 *ACS Publications*
[16] Lee K-S, El-Sayed MA 2005 *The Journal of Physical Chemistry B* **109** 20331-8
[17] Wiesner J, Wokaun A 1989 *Chemical physics letters* **157** 569-75
[18] Orendorff CJ, Murphy CJ 2006 *The Journal of Physical Chemistry* B **110** 3990-4
[19] Jain PK, Lee KS, El-Sayed IH, El-Sayed MA 2006 *The journal of physical chemistry B* **110** 7238-48
[20] Chen Y-S, Frey W, Kim S, Kruizinga P, Homan K, Emelianov S 2011 *Nano letters* **11** 348-54
[21] Chen Y-S, Frey W, Kim S, Homan K, Kruizinga P, Sokolov K, et al 2010 *Optics express* **18** 8867-78
[22] Luke GP, Bashyam A, Homan KA, Makhija S, Chen Y-S, Emelianov SY 2013 *Nanotechnology* **24** 455101
[23] Zijlstra P, Chon JW, Gu M 2009 *nature* **459** 410-3
[24] Awad H, Abdallah T, Negm S, Talaat H 2020 *MS&E* **762** 012004
[25] Zhu X-M, Fang C, Jia H, Huang Y, Cheng CH, Ko C-H, et al 2014 *CellNanoscale* **6** 11462-72
[26] Wang P, Wang X, Wang L, Hou X, Liu W, Chen C 2015 *Science and technology of advanced materials* **16** 034610
[27] Vetten MA, Tlotleng N, Rascher DT, Skepu A, Keter FK, Boodhia K, et al 2013 *Particle and fibre toxicology* **10** 50
[28] Chen L-C, Wei C-W, Souris JS, Cheng S-H, Chen C-T, Yang C-S, et al 2010 *Journal of Biomedical optics* **15** 016010

Effect of UV-activated TiO_2 Nanoparticles on the Properties and Performance of PAni-TiO_2 Nanocomposite Films for Solar Cell Applications

F. M. El-Hossary [1], A. Ghitas [2], A. M. Abd El-Rahman [3, 1], A. A. Ebnalwaled [4,5], M. Abdelhamid Shahat [2,*], M. H. Fawey [1]

[1] Physics Department, Faculty of Science, Sohag University, 82524 Sohag, Egypt.
[2] Photovoltaic Unit, Solar and Space Research Department, National Research Institute of Astronomy and Geophysics (NRIAG), 11421 Helwan, Cairo, Egypt.
[3] King AbdulAziz University, Jeddah, KSA.
[4] Electronics & Nano Devices Lab, Physics Department, Faculty of Science, South Valley University, 83523 Qena, Egypt.
[5] Egypt Nanotechnology Center (EGNC), Cairo University Sheikh Zayed Campus, 12588 Giza, Egypt.

* E-mail: m.abdelhamid999@gmail.com

Abstract. To improve the performance of organic solar cells by enhancing the properties of the photoactivated nanocomposite layer, the UV irradiation process was used to activate titanium dioxide nanoparticles (TiO_2 NPs). Herein, polymer solar cells were fabricated with FTO/(PAni-TiO_2)/Ag system. A series of mixed polyaniline (PAni) with 20% of activated TiO_2 NPs at different processing times was used to form PAni-TiO_2 nanocomposite films. The structural evolution, surface characteristics, optical and electrical properties of PAni-TiO_2 films have been investigated. XRD patterns showed that the UV treatment of TiO_2 NPs increased the crystallite from 18.35 to 24.1 nm and the degree of crystallinity increased by 5.6%. The irradiated PAni-TiO_2 films showed a rougher and more porous surface compared to the untreated one. Moreover, the adhesion force and electrical conductivity of the treated nanocomposite films at 8 h improved to be 137 mN/m and 6.62 S/m, respectively. Incorporation of activated TiO_2 NPs exposure to UV for different times from 0 to 8 h with the PAni matrix enhanced the current density (J_{sc}) of PAni-TiO_2 based nanocomposite solar cells from 3.11 to 4.83 (mA/cm^2) and their efficiency from 0.33 to 0.85%. The increase in the solar cell efficiency is mostly ascribed to a structural change accompanied by a rapid increase in surface roughness, which led to a decrease in the reflected photons and thus an increase in the charge carriers produced. These results revealed the effect of surface UV irradiation of TiO_2 NPs on their structural properties and the electronic contact between PAni and TiO_2 NPs, which greatly influenced the amount of carrier transport within the PAni-TiO_2 composites.

Keywords: Titanium dioxide (TiO_2), Ultraviolet (UV) Activation, PAni-TiO_2 Nanocomposite Films, Electrical Conductivity, Hybrid Solar Cells.

1. Introduction

Metal oxide nanoparticles (NPs) used as effective materials for electronic applications owing to their nanometric size that brings unique physical, chemical, and biological characteristics. Among these, TiO_2, ZnO, ZnS, GO, CdS and Fe_2O_3 NPs are widely used and offer extraordinary potential candidates to remove toxic elements [1]–[3].

In particular, NPs of titanium dioxide (TiO_2) have attractive properties including high surface area, simplicity of synthesis, non-toxicity, cheap cost, excellent chemical stability and distinctive optical, electrical, and magnetic properties.

These characteristics make TiO_2 NPs suitable for various applications, such as solar-photovoltaic, water splitting, electrochromic devices, pigment, cosmetics and photocatalysis [4]. In addition to that, it can be used for energy storage, sensors, optics, piezoelectric, and biotechnology [4], [5]. As it is known, optimizing the crystal size and phase structure of TiO_2 NPs leads to control of the optical energy gap for specific kinds of applications. The energy band gap has values of 3.0 and 3.13 eV for rutile and brookite structure, respectively [6]. However, it was found to be higher than 3.1-3.2 eV for the anatase structure and therefore, it could be excited by ultraviolet (UV) irradiation ($\lambda <$ 380 nm) [7]. Various chemical and physical methods have been utilized for the preparation of TiO_2 NPs such as solvothermal, sol-gel, sonochemical, hydrothermal, chemical vapor deposition (CVD), physical vapor deposition (PVD), electro-deposition, etc. [8]. Activation by UV irradiation is one of the promising methods that control the surface area, crystallinity, and morphology of TiO_2 nano-powder, which leading to enhance its physical properties [9]. For example, by applying high energy UV irradiation on anatase TiO_2 NPs enhances charge carrier transport or photoresponsivity. This is achieved through generating electron-hole pairs that in turn generate active free radical species, which can interact with the polymer matrix [9], [10].

The UV treatment has been employed to enhance the electrocatalytic activity of TiO_2 nanotubes. The UV treated TiO_2 nanotubes, revealed superior electrocatalytic activity that ascribed to the enhanced donor density and conductivity which in turn attributed to the formation of Ti^{3+} and oxygen vacancies [11]. Moreover, the UV treatment process is often performed to improve surface hydrophilicity and wettability [12]. Besides, UV irradiation has a great impact on the microstructure, wettability, optical behavior and electrical conductivity of TiO_2 thin films [4].

On the other hand, conjugated polymers are widely used in various applications due to their good physical, mechanical, chemical and electrical properties [13]. Unlike other conductive polymers such as polypyrrole and polythiophene, the conductivity of PAni varies from semiconductor to metal conductivity; and it can be easily controlled by modifying the mixing level [14]. Besides, PAni possesses many other attractive properties such as positive temperature coefficient of resistance (PTCR), photoelectric properties, easy and inexpensive to synthesize, multicolor electrochromic, and chemical sensitivity. Therefore, PAni can be used in numerous applications such as photovoltaic (PV) cells, gas sensors, anticorrosion coating, and tissue engineering [14], [15].

Recently, the inorganic materials composition and conducting polymers have great interests as merits that can be combined to achieve a final composite with good performance [16]. Previous studies have approved that the physical and chemical properties of PAni-TiO_2 nanocomposite differ completely from pure PAni and TiO_2, due to a strong chemical interaction between the components [16]–[18]. Unlike the individual components, PAni-TiO_2 nanocomposites exhibit high electrical conductivity properties [19]. Moreover, TiO_2 molecules can accumulate significantly in aprotic solvents; as a result, both the depth of light penetration and the surface area of TiO_2-polymer interface will be reduced.

As it is known, UV irradiation has been extensively widely used to activate TiO_2 NPs to for use as effective water purification catalysts. Herein, based on our knowledge, this is the first time to use UV irradiated TiO_2 NPs with improved electrical conductivity in synthesizing a high-performance nanocomposite layer for bulk-heterojunction (BHJ) organic solar cells. In order to evaluate the effect of using UV irradiated TiO_2 NPs on the properties of PAni-TiO_2 nanocomposite films, the microstructure, surface morphological analysis and properties of the synthesized films were considered and characterized. Further, the film optical and electrical properties were investigated and discussed. Accordingly, to assess the performance of PAni-TiO_2 nanocomposite layers that used different irradiated TiO_2 NPs, typical solar cells were fabricated. The J-V characteristic curves were measured in dark and under simulated sunlight with a total light intensity of 100 mW/cm^2 to estimate the solar cell efficiency.

2. Experimental procedures

2.1. Reagents and Materials

Titanium tetrachloride ($TiCl_4$), which supplied by Loba (purity: 99.5%), was used as raw materials for TiO_2 NPs preparation. Moreover, aniline monomer C_6H_7N (purity: +99%), hydrochloric acid HCl (concentration: 37%) and ammonium peroxydisulfate (APS) $(NH_4)_2S_2O_8$ were used to prepare PAni. All chemical materials were purchased from Sigma Aldrich.

2.2. Preparation of TiO_2 nanoparticles

30 mL of $TiCl_4$ was added to 300 mL of distilled water in a flask surrounded by an ice bath to prevent TiO formation and then, the mixture was placed on a magnetic stirrer (1000 rpm). The chemical reaction of the solution was kept at this condition for 1h then, the solution temperature raised to 70 °C for 2 h using a hotplate. After 3 days, the obtained suspension was washed several times with distilled water and thus, dried at a temperature of 70 °C. Finally, it was calcined at 450 °C for 3 h.

For UV treatment, the obtained TiO_2 NPs were activated under UV lamp at a wavelength of 364 nm and a working power of 15 W at different times of 2, 4, 6, 8, and 10 h. The distance between the TiO_2 NPs and the UV source 10 cm.

2.3. Synthesize of PAni-TiO_2 nanocomposites films

Chemical oxidative polymerization method was used to prepare PAni-TiO_2 nanocomposites, where the aniline monomer polymerized in the presence of HCl and ammonium persulfate (APS) as a catalyst and a strong oxidizing agent, respectively. APS is a colorless inorganic salt, highly soluble in water, which has a chemical formula of $(NH_4)_2S_2O_8$. The experimental method and reaction mechanism between the activated TiO_2 NPs and the aniline monomers for the formation of PAni-TiO_2 nanocomposites are illustrated in Fig. 1 and Fig. 2, respectively.

A 0.01 mol of aniline monomer is dissolved in a 0.01 mol of HCl acidic solution and mixed using an ultrasonic unit (Elmasonic S 40H) at room temperature. In another beaker, A 0.01 mol of APS was dissolved in 15 mL distilled water to prepare a dilute APS solution. To cover the TiO_2 NPs with conductive PAni as a shell, TiO_2 powder was dispersed in the beaker containing aniline monomer and placed on a magnetic stirrer for 30 min. TiO_2 solution was mixed to the aniline mixture with ratio 20% vs. 80% aniline solution followed by several minutes in the ultrasound to confirm the complete interaction between aniline and TiO_2 NPs. A dilute APS solution was placed into a burette and was slowly added dropwise to the mixture of aniline–HCl–TiO_2 under magnetic stirring (1000 rpm) in an ice bath at a reaction time of 5.5 h. Accordingly, the formation of PAni-TiO_2 nanocomposites can be performed using a modular physicochemical technique [20], which begins with a slow drop rate at approximately one drop/second and continues under stirring to ensure good oxidization and homogenous polymerization of the PAni solution. The color is observed to be changed to dark green at the end of physicochemical reaction. This indicates that an oxidation process is underway, which ends with the formation process composite solution of PAni-TiO_2 nanocomposites. Then, the precipitate was filtered and washed with distilled water and ethanol to remove excess acid. The final composites solutions were deposited by a spin coating process on FTO substrates at 800 rpm for 30 s followed by drying at 50 °C for 30 min. The deposition and desiccation processes were carried out in the air atmosphere. The same process was repeated by adding activated TiO_2 NPs at different exposure times (2, 4, 6, 8 and 10 h).

On the surface of TiO_2-NPs various hydroxyl groups are present. The reaction between H_2 bond and ethanol on the TiO_2 surface weakens the agglomeration of TiO_2. Then the aniline solution disperses the TiO_2 NPs properly. The chemical bond formed between TiO_2 and PAni during the polymerization process is stronger than the H_2 bond between the TiO_2 surface and the ethanol hydroxyl group [21].

IOP Conf. Series: Materials Science and Engineering **956** (2020) 012015 doi:10.1088/1757-899X/956/1/012015

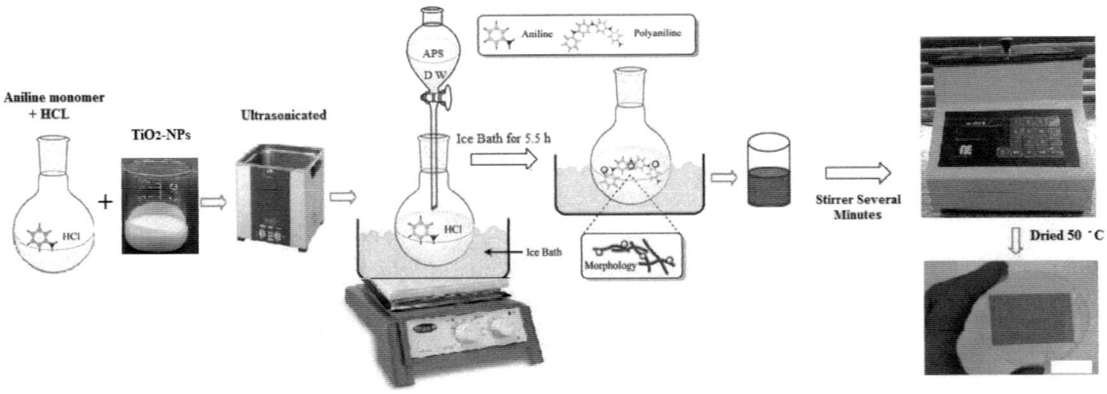

Figure 1. Experimental procedure for the synthesis of the Pani-TiO_2 nanocomposite

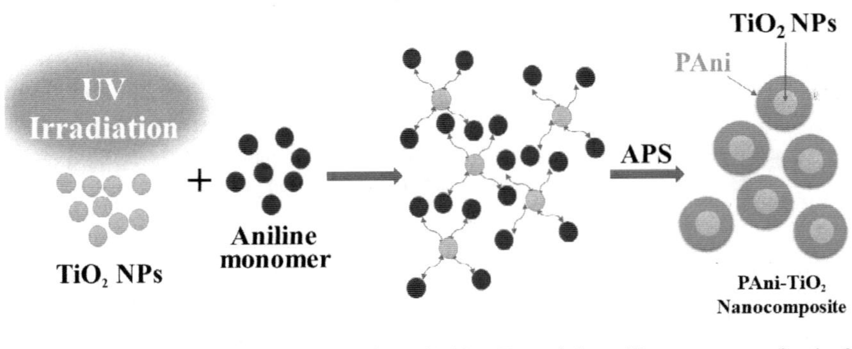

Figure 2. The reaction mechanism between the activated TiO_2 NPs and the aniline monomers for the formation of PAni-TiO_2 nanocomposites

2.4. Fabrication of FTO/PAni-TiO$_2$/Ag solar cell devices

Fluorine-doped tin oxide (FTO) (80-82% transmittance and 4 x 10^{-5} Ω.cm resistivity, sigma Aldrich) coated glass thin film was used in the fabricated organic solar cell devices as a transparent conductive anode. A metallic silver (Ag) paste (high concentration ≥75%, low resistivity of 1-3 x 10^{-5} Ω.cm) conductive material is used as a highly effective cathode. The FTO glass substrate is exposed to a special cleaning procedure using a dilute acidic solution of HCl followed by an extensive water rinse to remove the residual HCl. After that, it is sonicated in acetone from the surface of the conductive side, wet cleaned in ethanol and sonicate to remove organic impurities and finally rinse with deionized water. Then, the nanocomposites of PAni-TiO_2 as a photoactive layer is spin-coated on FTO surface. Finally, high quality silver bus bars as a cathode were printed on the top surface of the PAni-TiO_2 nanocomposite using a screen printer coater (EQ-SPC-2-LD). A schematic of the stack device architecture is shown in Fig. 3.

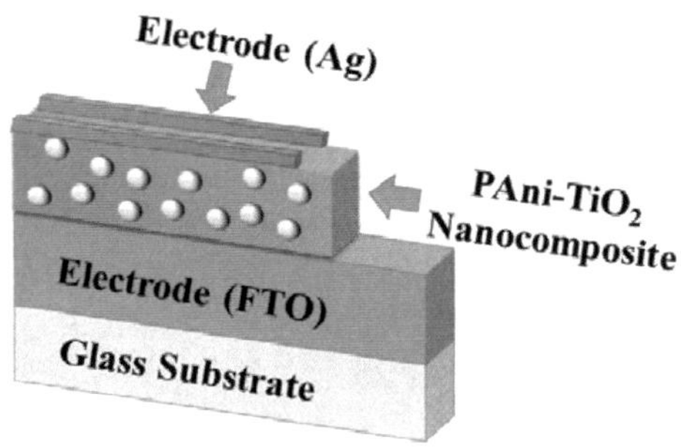

Figure 3. Organic solar cell architecture of FTO/PAni-TiO_2/Ag.

3. Characterization techniques

The X-ray Diffractometry (XRD) with a Cu target (Model D8 ADVANCE from Bruker) has been used to investigate the phase purity and structure of the prepared samples. The functional groups of TiO_2 NPs, pristine and treated nanocomposite thin films are evaluated using Fourier Transform Infrared (FTIR) Spectroscopy (Jasco Model 4100_Japan). The FTIR analysis is obtained from 4000 to 400 cm^{-1} with a resolution of 4 cm^{-1} at room temperature. The morphology of the obtained pure PAni and nanocomposite films was characterized by a high-resolution scanning electron microscope (SEM) (JOUL, JSM-5500, Japan) at an accelerated voltage of 10 KV. The optical properties of the nanocomposite films were evaluated using a computerized SPECORD 200 PLUS analytic Jena spectrophotometer with 1 nm step at normal incidence of light in the wavelength range of 190 – 1100 nm. An SEO Phoenix 300 contact angel analyzer was employed to measure the contact angle, wet energy, surface tension, diffusion of power factor and adhesion force. The surface roughness (Ra) was evaluated using a profilometer (Talysurf 50-Taylor Hopson precision). The electrical resistivity was measured with the four-point probe method using an EQ-JX2008-LD resistivity tester. The average electrical resistivity of each film was calculated from 10 reading sgained from various parts. Furthermore, the J –V characteristics of the solar cell under a fixed illumination intensity of 1.0 Sun at 25 °C is measured using a solar simulator system. The simulation test carried out in dark and illumination under standard condition (AM 1.5 G, 100 mW/cm^2) using a xenon lamp as a simulated light source irradiation with the use of a KEITHLEY 2400 Source Meter.

4. Results and Discussion

4.1. X-Ray Diffraction (XRD) Analysis

Fig. 4 shows typical XRD patterns of TiO_2 NPs, pure PAni, and PAni-TiO_2 nanocomposite films that treated by UV rays at different processing times, 2, 4, 6, 8 and 10 h, in compared to the untreated one. In all samples, peaks of diffraction intensities and widths are detected between 10° and 80° due to the parallel and perpendicular periodicity of the polymer (PAni) chain [22]. The structure of TiO_2 NPs primarily exists in three-phases: anatase, brookite, and rutile. Despite the rutile phase is highly persistent as bulk materials, preparation methods that include up to 450 °C are mostly produce in favor of anatase structure [23]. Anatase phase is the most active catalytic phase among all of the other TiO_2 phases, as its surface energy is lower compared to the rutile and brookite phases [23]. Herein, the diffraction peaks of TiO_2 are in a good agreement with the tetragonal TiO_2 phase with anatase crystal phase (a = b = 3.7335 Å, c = 9.3788 Å, $\alpha = \beta = \gamma = 90.0$) (ICSD, Code no 9852) [24]. Moreover, the diffraction peaks of the pure PAni are in a good agreement with the orthorhombic PAni structure (a = 7.3574 Å, b = 6.9675 Å, c = 10.1714 Å, $\alpha = \beta = \gamma = 90.0$) [25], [26].

The diffraction peak at 2θ of 23° is due to polymer chains placed parallel and closed to each other producing crystalline zones surrounded by amorphous areas [27]. The polycrystalline diffractions of the untreated and UV-treated PAni-TiO_2 nanocomposite films are identical to those caused by the standard powder diffractions of tetragonal anatase TiO_2 structure and orthorhombic PAni. This indicates that no phase transformation involved in the UV activation process. Further, the peaks intensity of the nanocomposite films is strongly influenced by UV radiation, where it gradually decreases with increasing the processing times. These results may be attributed to the improvement in the physicochemical interaction between the activated-TiO_2 NPs and PAni with an increase in the UV irradiation time. Besides, the peaks corresponding to the planes (211), (221) and (201) are slightly shifted towards a lower degree of 2θ. These little changes revealed that UV radiation does not affect the lattice structure of PAni-TiO_2 nanocomposite films. Moreover, by increasing the processing times, (323) PAni peak and (323), (103), (112), (213), (204) and (116) TiO_2 peaks have gradually vanished from the patterns of the treated PAni-TiO_2 films that attributed to the restrictive effect of the surface of TiO_2 nanoparticles which increases under UV light [7].

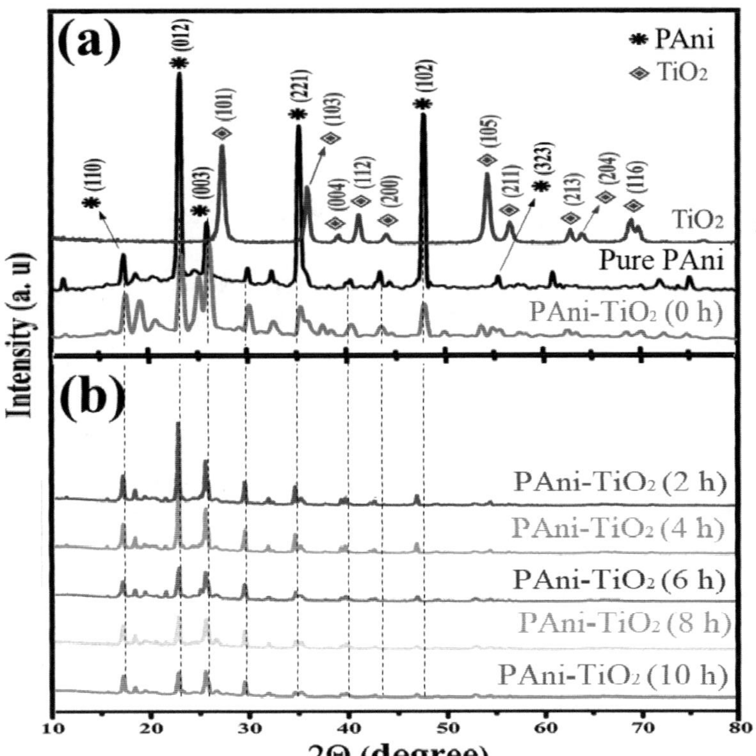

Figure 4. (a) XRD patterns of TiO_2 NPs, pure PAni and untreated PAni-TiO_2 nanocomposites films, (b) treated PAni-TiO_2 nanocomposites films synthesized of pure PAni with 20% of TiO_2 powder treated under UV light at processing time from 2, 4, 6, 8 and 10 h.

In general, crystallinity depends on the periodicity, that is the long-range order of atoms, ions or molecules in a particular sample. XRD patterns demonstrate that UV treatment has improved the crystallization of nanofilms as shown in Table 1. The XRD studies revealed an increase in the crystallinity of PAni-TiO_2 treated films compared to the untreated one. The crystallinity increased from 79.7% for the untreated film to 86.2% for one treated at 10 h processing time.

The crystallite size (D) can be calculated from the average of all main peaks using the Debye–Scherrer equation [28]:

$$D = \frac{0.94\,\lambda}{\beta\,cos(\theta)} \quad (1)$$

Where λ is the wavelength, β is the full width at half maximum of the peak, and θ is the diffraction angle. The crystallinity and the crystallite size of pure-TiO_2 NPs are 80.6% and 18.35 nm, respectively, as shown in Table 1 and gradually increased as the duration of exposure to UV compared to pristine PAni-TiO_2 nanocomposite. This can be attributed to the UV radiation that strongly affects the TiO_2 surface leading to different interactions between PAni and TiO_2 in the nanocomposite films and also attributed to the physical dependence of the polymer on the TiO_2 NPs surface.

Table 1. Crystallite size and crystallinity of UV treated and untreated PAni-TiO_2 nanocomposite samples

Sample	Crystallite size (nm)	Crystallinity (%)
TiO_2 powder	18.35	80.6
Pure PAni	22.39	77.3
PAni-TiO_2 (0h)	20.21 – 2	79.7
PAni-TiO_2 (2h)	21.51	83.4
PAni-TiO_2 (4h)	22.20	83.5
PAni-TiO_2 (6h)	23.42	83.7
PAni-TiO_2 (8h)	23.76	84.5
PAni-TiO_2 (10h)	24.10	86.2

4.2. FTIR Spectra

FTIR spectra was obtained in order to confirm the interaction formulation of the chemical compounds that formed between TiO_2 NPs, pure PAni to form PAni-TiO_2 nanocomposite as seen in Fig. 5. Three main absorption bands are observed in TiO_2 NPs sample. The first peak is appeared at 3437 cm^{-1} and it belongs to stretching vibrations hydroxyl (O–H), representing the water as moisture. The second peak is observed at 1627 cm^{-1}, and it belongs to stretching Ti–O bond [29]. The third one in a broad and strong peak between 800–450 cm^{-1} is assigned to vibrations in the stretching Ti–O–Ti bonds [30], [31]. In case of the PAni-TiO_2 nanocomposite, the main peaks of pure PAni bands continue to appear clearly in a nanocomposite in the region 3139–472 cm^{-1} with intensity changes and a small peak shift towards higher wavenumbers, which assigned to the presence of TiO_2-NPs in the nanocomposite structure. The absorption peak between 3690–2983 cm^{-1} can be attributed to the stretching of O–H and hydrogen-bond N–H stretching of aromatic amine [32], [33]. Surfaces of the hydroxyl bonds can accept photo-induced holes and form hydroxyl radicals [34]. As a result, these radicals can improve the photocatalytic activity of the composite [35]. The characteristic peaks at 1537 cm^{-1} and 1412 cm^{-1} are assigned to C=N and C=C ring asymmetric and symmetric stretching vibrations, due to the presence of quinonoid structures and benzenoid ring, respectively. The peak at 1301 cm^{-1} corresponds to C–N stretching of the benzenoid unit. The peak at 1193 cm^{-1} is due to the quinonoid unit of the PAni doped, which is used to measure the degree of delocalization and is characterized as the peak of the PAni conductivity [36]. The existence of the main peaks of PAni (i.e. C=C and C–N peaks) in the spectrum of PAni-TiO_2 nanocomposite confirmed the formation of PAni on nano TiO_2 surface [7]. The peaks at 870-743 cm^{-1} are associated with C–C and C–H in the benzenoid unit [20]. Whereas, the peak at 616 cm^{-1} is relatively enhanced compared to pure

PAni, which may be attributed to the out-of-plane bending flexural vibration of C—H aromatic rings of PAni and vibrations in the stretching Ti—O—Ti bonds of TiO_2 [32].

According to the results, the peak appears at 3193 cm^{-1} in a nanocomposite characterized by high intensity, wide stretching shape and small shifting towards short wavenumber compared to that found in pure PAni. This might be attributed to the attraction by a group of adjacent atoms or two nearby functional groups [37]. Moreover, the decrease in the intensity of the characteristic peak at 616 cm^{-1} for TiO_2-NPs compared to the corresponding finger peak in the PAni-TiO_2 nanocomposite indicates a structural change in the polymer matrix resulting from a strong H-bonding interaction exists between PAni and TiO_2-NPs. The good interaction of PAni to TiO_2-NPs surface led to shield the nanoparticles with a coating PAni layer due to coordination between the N atoms of the PAni chains and the TiO_2 NPs [38], [39].

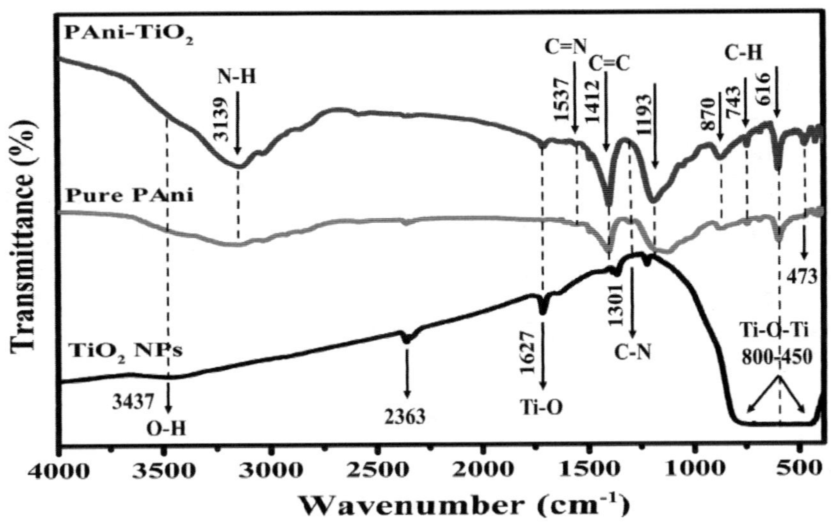

Figure 5. FTIR spectra of TiO_2-NPs, pure PAni and PAni-TiO_2 nanocomposite

4.3. Morphology

SEM images of pristine TiO_2, pure PAni, untreated and treated PAni-TiO_2 composite films are shown in Fig. 6. From the SEM micrographs, a porous topography of pristine TiO_2 and pure PAni is observable. Based on the previous literature, the increased active layer with a microporous structure results in a relatively rough surface, which increases the surface contact between the nanocomposite layers and the electrode surface [40]. In the microporous surface of the active layer of the composite, the increased capacitance can take place as a result of maximum absorption [41]. Moreover, pristine and activated TiO_2 NPs are dispersed in a PAni matrix showing a thick film integrated with featured structure and non-agglomerated TiO_2 NPs [27], [42]. This could be attributed to the formation of PAni shell on the surface of TiO_2 NPs that causes repulsion forces between nanoparticles preventing their agglomeration [41]. The morphology of the nanocomposite particles depends on the conditions of polymerization such as pH, temperature, polymerization time as well as the TiO_2 structure and ratio [43]. Besides, the white area is assigned to be TiO_2, while the grey and black areas refer to the polymer chain. By increasing the processing time from 0 to 8 h (Fig. 6b-g), the activated PAni-TiO_2 films showed improvement in the white surface area distribution compared to the untreated one. This may confirm that the activated TiO_2 is distributed uniformly over the surface of the polymer molecules with increased activation time. Furthermore, the formation of nanoflakes in the nanocomposite structures was observed for samples containing TiO_2 NPs that were exposed to UV radiation for a relatively long period of time 8 and 10 h (see

Fig. 6g and 6h) [40], [41]. This morphological phenomenon can be attributed to the improvement in the physicochemical reaction between the activated TiO_2 and PAni.

Increasing the processing times results in a significant improvement in the crystallization of the nanoparticles and reactive surface area with PAni chains. Moreover, the agglomerating aspect vanishes as activated-TiO_2 NPs, and well distributed on the long PAni chains; causing a repulsive action preventing agglomeration process. This leads to creating a morphological structure with nanoflakes. As the UV irradiation times of TiO_2 NPs increase, the adsorptive properties of the PAni matrix increase and the surface area for nanoflakes increases. These nanoflakes improve the efficiency of electrical properties and photocatalytic activity [21].

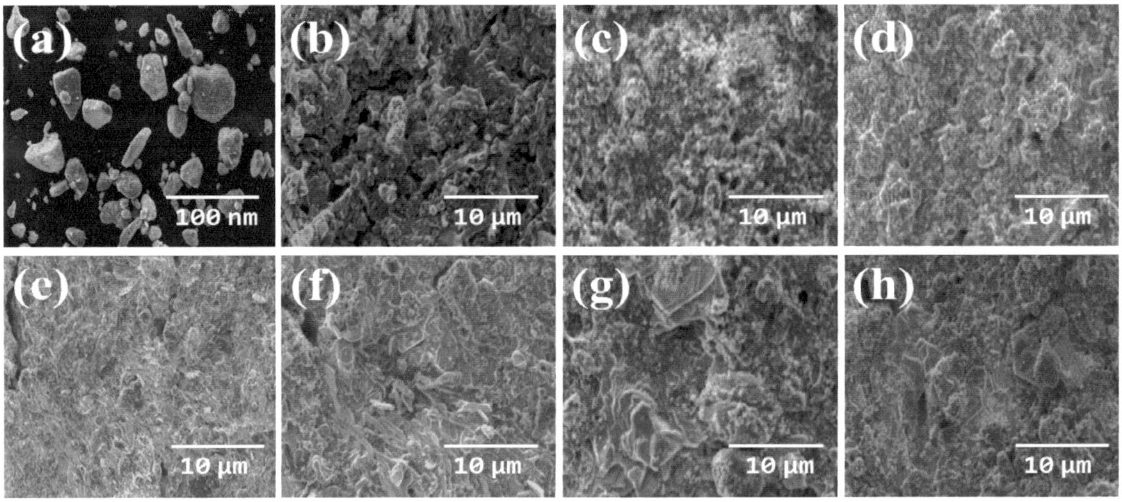

Figure 6a-h. SEM images of a) Pristine TiO_2 NPs, b) Pure PAni film, c) Untreated PAni-TiO_2 nanocomposite film, d:h) PAni-TiO_2 nanocomposite films of treated TiO_2 powder by UV at 2, 4, 6, 8 and 10 h, respectively, at fixed mixing ratio of 20% TiO_2 NPs.

4.4. Optical Properties

The optical properties of pure PAni film and PAni-activated TiO_2 nanocomposite films as a function of UV treatment time has been investigated and discussed. Fig. 7a,b reveals the absorbance and transmittance spectra of all films at near-normal incidence over a wavelength range of 300–800 nm. Fig. 7a shows a similar spectrum of broad absorption peaks in the range of 325–465 nm and 550–800 nm for pure PAni and nanocomposite films with absorption peak shifts. The absorption band of 550–800 nm shifts towards lower wavelengths with increasing the UV irradiation time of the TiO_2 NPs up to 8 h. This band shift indicates the effect of activated TiO_2-NPs on electronic structure of TiO_2 NPs in valence band that could interact with the higher occupied molecular orbital (HOMO) of the PAni [21]. On the other hand, increasing the processing time to 10 h leads to a different behavior where a small shift towards the higher wavelengths occurs. These shifts indicate the formation of nanocomposite with TiO_2 [44]. Pure PAni shows as an energy absorption equivalent at wavelengths of 330, 420 and 600 – 800 nm which in turn leads to transitions of $\pi \rightarrow \pi^*$ within the benzenoid rings, Polaron$\rightarrow \pi^*$ of quinoid rings, and $\pi \rightarrow$polaron excitonic of the quinoid rings, respectively [45]. These observed peaks gradually increase in intensity with a slight shift towards higher wavelengths in the PAni-TiO_2 nanocomposites. This confirms the formation of the polaron tape in the bandgap of the PAni and was strongly influenced by the incorporation of treated TiO_2 NPs. A peak is also observed around 690 nm for pure PAni which correlates to the benzenoid and quinoid excitonic transition.

A minimum absorption peak is mostly observed for untreated PAni-TiO_2 nanocomposite (at 0 h). Thereafter, absorption was gradually increased as the UV radiation treatment times increased to reach the optimum value at 8 h of treatment. This can be attributed to the good interactions between PAni and TiO_2 NPs [46]. The same behavior for PAni-TiO_2 absorption curve has been previously observed [47], [48]. It is well known that PAni-TiO_2 nanocomposites display applications on a large scale for photocatalytic when they have good absorption in the ultraviolet spectrum range [47].

In general, the transmittance spectra of PAni-TiO_2 nanocomposite films are strongly different depending on the UV light treatment time. The maximum transmission has been detected at the pristine TiO_2 NPs and, hence, gradually decreased with increasing the treatment time to have the lowest value at a treatment time of 8 h. This corresponds to π-π* transition in the UV and bipolaronic band in the visible regions.

The optical band gap *(E_g)* of the samples can be calculated using the universal method, Tauc relationship, for direct allowed transition semiconductor materials [49].

$$\alpha h\nu = B\left(h\nu - E_g\right)^{1/2} \qquad (2)$$

Where $h\nu$ is photon energy and B is a parameter which is related to the transition probability, α is the absorption coefficient and E_g is the optical band gap [50]. The E_g values are directly related to the film composition and structure. On the other hand, it has been found that its value increased with increasing the time of exposure to UV rays of TiO_2 powder as it is listed in Table 2. These values are higher than that calculated value for the unmodified PAni-TiO_2 (1.65 eV). The optical absorption has found to be dependent on the short-range order in the amorphous state and defect state [51]. In the present system, the increased of the optical band gap might be attributed to the reduction of the system disorder and the increase of the defect state density. Tone can conclude that introducing treated TiO_2 into the PAni has a potential effect on the optical properties of active nanocomposite films.

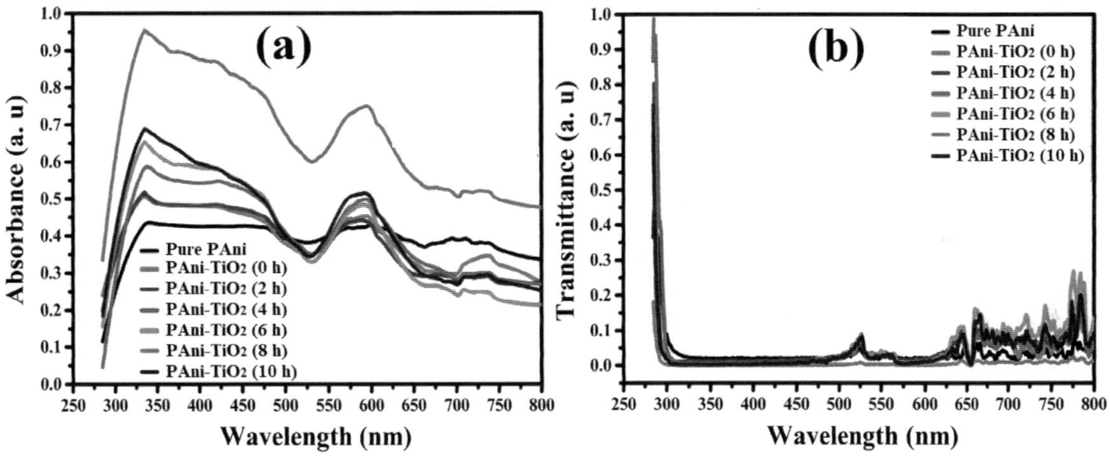

Figure 7a,b. a) absorbance, b) transmittance for pure PAni, pristine and modified PAni-TiO_2 nanocomposites using UV irradiation for TiO_2 powder at various times.

Table 2. Energy band gap values of the obtained nanocomposite films treated at different times

Treatment Time (h)	Pure PAni	PAni-TiO_2 (0 h)	PAni-TiO_2 (2 h)	PAni-TiO_2 (4 h)	PAni-TiO_2 (6 h)	PAni-TiO_2 (8 h)	PAni-TiO_2 (10 h)
Eg (eV)	2.26	1.65	1.77	1.76	1.87	1.99	1.65

4.5. Water droplet interaction with PAni-TiO_2 nanocomposites

The contact angles (CAs), adhesion force, wettability, and water droplet spreading coefficient on the surface of PAni-TiO_2 nanocomposite films were strongly affected by the activation degree of TiO_2 using UV rays as shown in the Fig. 8a-d. The CA of the untreated PAni-TiO_2 film was approximately 52.81°, which gradually decreased to be 46.07°, 36.02°, 32.96°, 27.46° after UV treatment time of 2, 4, 6 and 8 h, respectively. This indicates that the PAni-TiO_2 nanocomposite films are highly hydrophilic. Then, the CA value increased to 31.39° after UV treatment time of 10 h as shown in Fig. 8a. The fundamental explanations of the observed difference confirm the great influence of the crystallite size, molecular interactions [51], [52], and vertical topographical variation (roughness) on the electrochemical properties of PAni-TiO_2 surfaces [53], [54]. While the CA value was gradually decreasing, the work of adhesion was increased by increasing the treatment time to achieve the optimum value of 137.39 mN/m at a treatment time of 8 h, as shown in Fig. 8b. The work of adhesion
presents the adhesion force between obtained films and droplet water, and it can be calculated from the next equation [55]:

$$W_{sl} = \gamma_s + \gamma_l + \gamma_{sl} \quad (3)$$

where γ_s, γ_l, and γ_{sl} are the surface free energies of solid, liquid (water in this study), and solid-liquid interfacial energy, respectively. Combining this equation with the next equation of Young-Dupré

$$\gamma_s = \gamma_{sl} + \gamma_l \cos \theta_e \quad (4)$$

We got [56]:

$$W_{sl} = \gamma_l (1 + \cos \theta_e) \quad (5)$$

where θ_e is the equilibrium (Young's) contact angle between PAni-TiO_2 nanocomposite films and water, and γ_l is the water surface tension on the solid surface.

An examination of the interaction of a water drop with the modified TiO_2 nanostructured surfaces within PAni chains reveals that the water absorption on the nano-surface depends on the density of the surface charge, which is strongly affected by the nanoscale roughness and the surface physicochemical properties (among them wettability) [57]. The wettability and spreading coefficient of PAni-TiO_2 films showed that they are highly dependent on the modification time and the drop retraction due to the hysteresis of contact angles from rough surfaces, as seen in Fig. 8c,d [58]. The same behaviour was reported on PAni-TiO_2 nanocomposite films [59].

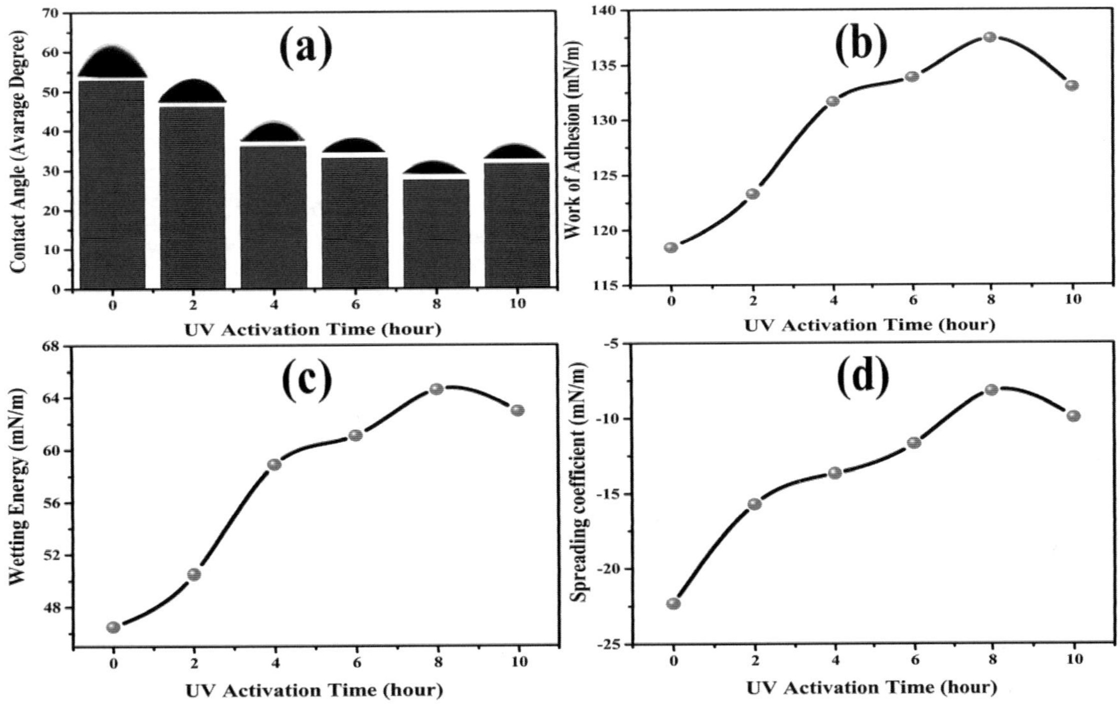

Figure 8a-d. a): Contact angle, b) adhesion force, c) wetting energy, d) spreading coefficient for pristine and modified PAni-TiO_2 nanocomposites using UV irradiation for TiO_2 powder at various times.

4.6. Roughness Measurements of PAni-TiO_2 Nanocomposite Films

Roughness measurements of nanofilms, consisting of the PAni matrix and activated TiO_2 NPs at different times under a UV lamp, revealed a great diversity. Fig. 9 shows the measured factors of surface roughness, i.e. average roughness (Ra) and root mean surface roughness (Rq). It was found that Ra and Rq have values of 2.74 and 3.72 μm, respectively for the untreated film. These values increased to the maximum values of 5.04 and 6.38 μm, for Ra and Rq, respectively with increasing the activation time to 8 h. Then, it decreased to 3.01 and 3.9 μm after 10 h of UV treatment time. The interpretation of this dissimilar trend is mostly ascribed to the difference in the structural features of both nanocomposites. The effect of molecular interactions, crystallite size of the irradiated-TiO_2 NPs within PAni matrix and their distribution over the long PAni chains leads to an increased surface roughness [51], [60]. However, the fibrous network structure is mostly responsible for surface smoothness. It fully agrees with the contact angle results. The change in the surface roughness of the nanocomposite films affects their optical properties including transmittance, absorbance and optical constants [61].

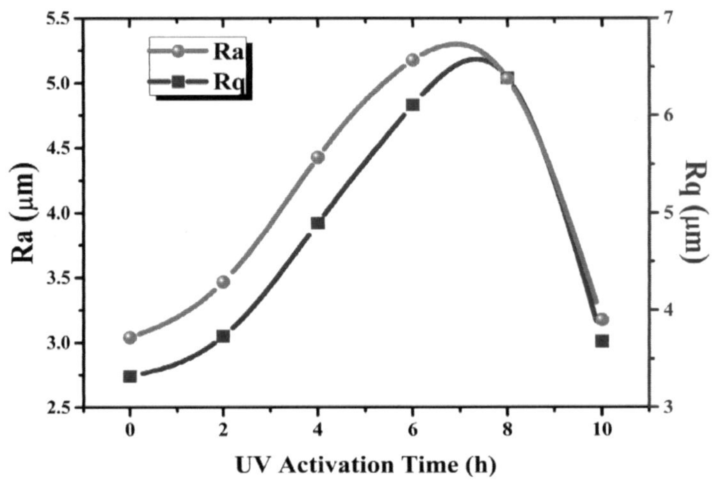

Figure 9. The surface roughness parameters for pristine and modified PAni-TiO_2 nanocomposites using UV irradiation for TiO_2 powder at various times.

4.7. Electrical Conductivity of PAni-TiO_2 Nanocomposite Films

The prolonged UV-activation process can enhance the catalytic and electrical properties of TiO_2 through holes induced by ultraviolet radiation and reducing the adsorbed oxygen by electrons, equation (6) [62].

$$\mathbf{TiO_2 + h\nu \rightarrow TiO_2\ (h^+_{V.B} + e^-_{CB})} \qquad \textbf{(6)}$$

Subsequently, PAni electrical properties can be enhanced by adding activated TiO_2 NPs inside the PAni chains [63]. Interaction between O^- ions on the surface of TiO_2 grains and the nitrogen atoms in the large PAni molecules causes electrons trapping by releasing the oxygen ions and returning to the surface of the grains [64].

The electrical properties of the obtained pure PAni and PAni-TiO_2 samples were investigated. As seen in Table 3, the conductivity of the unmodified PAni and treated PAni-TiO_2 at different treatment times of 0, 2, 4, 6, 8 and 10 h under UV irradiation. Resistivity is measured by the 4PP (EQ-JX2008-LD) method and approved that; UV irradiation treatment has a significant effect on the electrical conductivity of the obtained films. As observed from the current results, the electrical conductivity of pure PAni film is 4.77 S/m and increases to 5.26 S/m in the untreated nanocomposite film. However, for the synthesized nanocomposite films based on the UV activated TiO_2 NPs at different times (2, 4, 6 and 8 h) leads to gradually enhance the conductivity of films to 6.11, 6.32 and 6.54 and 6.62 S/m, respectively. This behavior has been previously investigated by others and has been mainly ascribed to the physicochemical reactions of activated-TiO_2 NPs within the matrix structure; which leads to a change in the oxidation state [65], [66]. It is known that the difference in oxidation level of PAni inevitably affects its electronic structure. Accordingly, the polaron band has found to be smaller for pure PAni (corresponding to absorption spectra) than that of phase PAni in the pristine composites, and it was gradually improved by increasing the UV processing times of TiO_2 NPs [65]. This in turn leads to high mobility of charge carriers (polarons) along PAni chains [66]. These show higher values than that of previous studies such as SU Bitao et al (2.86×10^{-7} S/m) [63] and Xingwei Li et al. (2.9×10^{-7} S/m) [67]. This significant improvement in conductivity may arise due to the transfer of electrons from the surfaces of the polar O^{2-} TiO_2 NPs to the PAni chains, causing them to be reduced. In a highly reduced PAni, localized charges can be strongly obtained in a PAni ring [64].

A similar effect was obtained on modified TiO_2 films using UV irradiation and correlated with an increase of n-type conductivity due to the formation of lattice or electronic defects, like oxygen vacancies or interstitial Ti^{4+} ions [4], [61]. These results prove that focusing the UV rays on TiO_2 NPs surfaces, may realize more carries and electronic interaction between PAni and TiO_2 in PAni-TiO_2 composites that may increase electrical conductivity. Furthermore, increasing the activation process to 10 h decreased the conductivity value to be 5.88 S/m. This may be attributed to a relative increase in the segregation of PAni chains upon prolonged treatment, with a high agglomeration of TiO_2 inside the PAni matrix. This results in the formation of a non-conducting agglomeration zone in the PAni-TiO_2 layer, which in turn reduces carrier mobility and the electrical conductivity [62]. Also, many factors such as the mixing ratio of TiO_2 NPs within the polymer matrix may, increase the reaction of Pani-TiO_2 and reduction of the conjugated lengths in the PAni chains [57,58], and may also occur due to a fundamental network imbalance or TiO_2 nano purity [59], [64].

Table 3. Electrical conductivity behavior for pure PAni, untreated nanocomposite film and treated PAni-TiO_2 nanocomposite films using UV irradiation for TiO_2 NPs at various times

Treatment Time (h)	**Pure PAni**	**PAni-TiO_2 (0 h)**	**PAni-TiO_2 (2 h)**	**PAni-TiO_2 (4 h)**	**PAni-TiO_2 (6 h)**	**PAni-TiO_2 (8 h)**	**PAni-TiO_2 (10 h)**
Conductivity (S/m)	4.77	5.26	6.11	6.32	6.54	6.62	5.88

4.8. Performance evaluation of solar cells based on PAni-TiO_2 nanocomposite films

Current density–voltage (J–V) measurements were obtained for the fabricated solar cells in the dark and illumination states. I–V tester device was used in the presence of a simulated solar spectrum applying 1000 W/m^2 with standard conditions including air mass (AM) of 1.5 G and temperature of 25 °C. The current flows in the solar cell circuit and recorded as a function of applied voltage under the stimulated white- light illumination. Herein, PAni's hybrid photoactive layers were designed as a donor mixed with activated-TiO_2 NPs as acceptor in the fabricated solar cells. As observed in the J–V characteristic in dark, the test devices that were structured by pure PAni, pristine and treated photoactive layers exhibited the same asymmetric diode-like behavior as shown in Fig. 10.

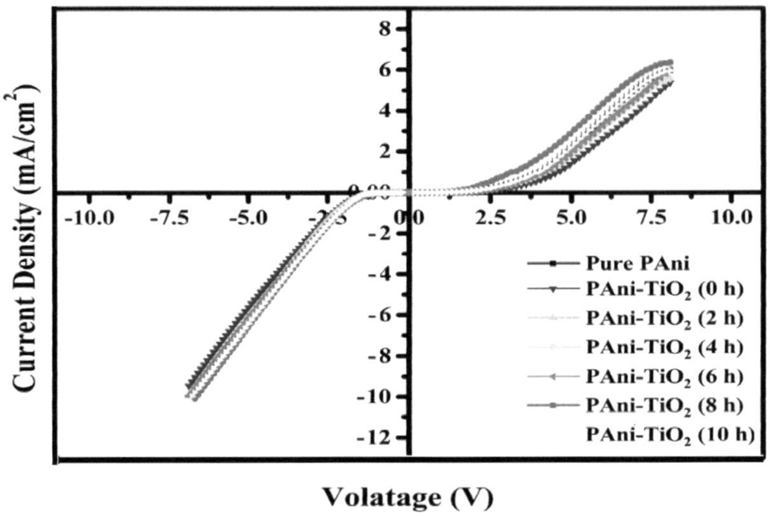

Figure 10. Diode-like behaviour of J-V curves obtained in the dark of designed photovoltaics based on pure PAni film, untreated nanocomposite film (pristine) and treated PAni-TiO_2 nanocomposite films using UV irradiation for TiO_2 NPs at different times (2, 4, 6, 8 and 10 h) as activate layers.

Incorporation of TiO_2 NPs within the PAni matrix increases the conductivity due to the contribution of polarons and bipolarons, which can move within shorter distances along polymer chains. Fig. 11 shows J–V curves of fabricated photovoltaic cells based on pure PAni, pristine composite and treated PAni-TiO_2 nanocomposites under illumination. As observed, the J–V curve is shifted towards higher values of current densities for the pristine composite device structured by FTO/PAni-TiO_2 (0h)/Ag compared to that one fabricated with FTO/pure PAni/Ag. These results indicate that incorporating of TiO_2 NPs within the PAni matrix significantly increases the carriers transport in the PAni structure by contributing to polarons and bipolarons, which move at shorter distances in polymer chains [68]. Moreover, incorporation of UV activated TiO_2 NPs at different processing times into the PAni-structure to form PAni-TiO_2 photoactive layer results in a gradual increase in current density values (J–V curves) compared to those designed with the pristine layer. The possible reason of the higher current density may be ascribed to (i) the increase in the physicochemical reactions between PAni and TiO_2 as well as the number of charge carriers that occurred by the activation process of TiO_2 NPs using UV irradiation at long processing times [69] (ii) a reduction in the interfacial resistance between TiO_2 and PAni owing to the increase in the number of polarons in PAni structure and the structural homogeneity of nanocomposites [70].

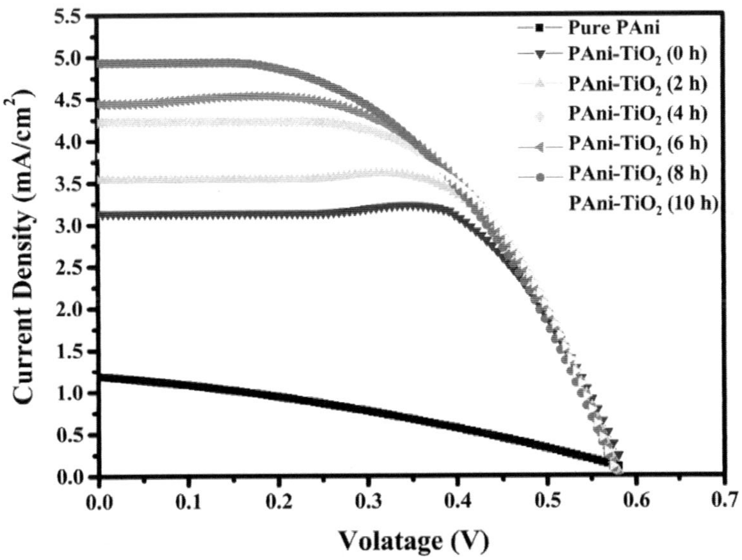

Figure 11. Illumination J-V characteristic curves of fabricated photovoltaic cells based on pure PAni, pristine composite and treated PAni-TiO_2 nanocomposites under simulated light irradiation 100 mW/cm^2 and AM 1.5 G.

Photovoltaic parameters of the solar cell device such as short circuit current density (J_{SC}), open circuit voltage (V_{OC}), fill factor (FF) and power conversion efficiency (η) can be attained from J–V characterization curves, as listed in Table 4. The V_{OC} is the voltage at the intercept of the J–V curve with the abscissa (horizontal axis) of the J–V diagram; whereas the J_{SC} is the value of current density at the intercept of the J–V curve with the ordinate (vertical axis). As observed, V_{OC} depends mostly on the trade-off energy between highest occupied molecular orbital (HOMO) of the donor and lowest unoccupied molecular orbital (LUMO) of the acceptor.

Herein, measurements of V_{OC} showed nearly constant values as the photoactive material (PAni) and have constant value (around 0.58 V) for all designed photovoltaic devices [71]. Moreover, it is known that the increase in J_{sc} relies mainly on the carrier's movement which includes generation, separations, transportation and collection on electrodes [72]. The photoactive layer modification process with UV-activated TiO_2 NPs plays an important role in improving short circuit current properties.

The photovoltaics efficiency of FTO/Pure PAni/Ag is 0.23% and increased to 0.33% by incorporating the pristine TiO_2 NPs into PAni structure. This is due to an increase of photocurrent density within composites structure [70]. The designed solar cells using activated-TiO_2 NPs exhibit a gradual improvement in the efficiency values of 0.42%, 0.55% and 0.69% at UV processing times of 2, 4 and 6 h, respectively. Then, it is increased to an optimum value of 0.85% for activated nanocomposite layer synthesized with a UV irradiated TiO_2 NPs for 8 h. All these manufactured devices show a better Performance based on their calculated efficiencies compared to that of other similar devices fabricated using of PAni-TiO_2 as active layers [69], [73], [74]. This performance improvement might be ascribed to the following reasons:

(i) At relatively high UV processing times, the physiochemical interaction between p-type PAni and n-type anatase-TiO_2 NPs can be increased and lead to an improvement in the donor/ acceptor interface that creates easy exciton dissociation [69].

(ii) The increase in the number of polarons and bipolarons in the PAni structure, which creates an efficient charge carries transfer network in the nanocomposite material. This interprets the conduction mechanism of the PAni-TiO_2 nanocomposite [75].

(iii) The rapid increase in surface roughness decreases the reflected photons and thus increases the produced charge carriers.

On the other side, prolonged processing time up to 10 h for TiO_2 NPs led to reduce the current density and cell efficiency to 3.81 (mA/cm^2) and 0.44%, respectively. This can be attributed to the reduction of exciton dissociation

resulted uneven distribution of TiO_2-NPs in PAni matrix along with uncontrolled aggregation for TiO_2 which can reduce the hole transportation [62].

Table 4. Characteristic parameters for pure PAni, pristine and modified PAni-TiO_2 nanocomposites solar cells

Sample	V_{oc} (V)	J_{sc} (mA/cm^2)	FF (%)	η (%)
FTO/Pure PAni/Ag	0.58	1.17	34.8	0.23
FTO/PAni-TiO_2 (0h)/Ag	0.58	3.11	18.0	0.33
FTO/PAni-TiO_2 (2h)/Ag	0.58	3.58	20.2	0.42
FTO/PAni-TiO_2 (4h)/Ag	0.58	4.17	20.8	0.50
FTO/PAni-TiO_2 (6h)/Ag	0.58	4.45	26.3	0.69
FTO/PAni-TiO_2 (8h)/Ag	0.58	4.83	30.3	0.85
FTO/PAni-TiO_2 (10h)/Ag	0.58	3.81	20.1	0.44

5. Summary and conclusion

TiO_2 NPs with anatase structure phase was successfully synthesized and then activated under UV irradiation at different exposure times for 2, 4, 6, 8 and 10 h. Activated-TiO_2 NPs was added during polymerization process of aniline monomers to synthesize highly conductive PAni-TiO_2 nanocomposites. Moreover, a PAni-TiO_2 nanocomposite solution was successfully obtained in a film form. The optical and electrical properties of the photoactivated PAni-TiO_2 nanocomposite films have found to be improved with increasing the processing times of the TiO_2 NPs. Further, the surface roughness and the porosity of the PAni-TiO_2 films were also increased. At UV irradiation time for TiO_2 NPs at 8 h, the crystallinity, optical energy gap and adhesion force of the PAni-TiO_2 nanocomposites were enhanced and found to be 86 %, 1.96 eV and 137 mN/m, respectively. Furthermore, the electrical conductivity of PAni-TiO_2 films was improved to 6.623 S/m after a UV treatment time of 8 h, where the transport of charge carriers in PAni-TiO_2 composites was increased. The current density (J_{sc}) of the PAni-TiO_2 nanocomposite based solar cells was increased from 3.11 to 4.83 (mA/cm^2) and the efficiency increased from 0.33 to 0.85% as the UV processing times of TiO_2 NPs increased from 0 to 8 h. This performance improvement may be ascribed to a structural effect along with the rapid increase in surface roughness which can decrease the reflected photons and thus increase the produced charge carriers.

References

[1] Lizama C, Freer J, Baeza J and Mansilla H D 2002 *Catal Today* **76** 235–246.
[2] Poulios I, Micropoulou E, Panou R and Kostopoulou E 2003 *Appl Catal B Environ* **41** (4) 345–355.
[3] El-Hossary FM, Ghitas A, Abd El-Rahman AM, Ebnalwaled AA, Fawey MH and Shahat MA 2020 *IOP Conference Series: Materials Science and Engineering* **762** 012001.

[4] Addonizio M L, Aronne A and Imparato C 2020 *Appl Surf Sci* **502** 144095.
[5] Hagfeldt A and Graetzel M 1995 *Chem Rev* **95** 49–68.
[6] Reyes-Coronado D, Rodríguez-Gattorno G, Espinosa-Pesqueira M E, Cab C, De Coss R d, Oskam G 2008 *Nanotechnology* **19** (14) 145605.
[7] Monfared A H, Jamshidi M, Hashemi Monfared A, Jamshidi M, Monfared A H and Jamshidi M 2019 *Prog Org Coatings* **136** 105257.
[8] Manjunath K, Yadav LSR, Jayalakshmi T, Reddy V, Rajanaika H, Nagaraju G, et al 2018 *J Mater Res Technol* **7** (1) 7–13.
[9] Zenteno A, Lieberwirth I, Catalina F, Corrales T, Guerrero S, Vasco DA, et al 2017 *Compos Part B Eng* **112** 66–73.
[10] Zapata PA, Rabagliati F M, Lieberwirth I, Catalina F, Corrales T 2014 *Polym Degrad Stab* **109** 106–114.
[11] Tian M, Thind S S, Dondapati J S, Li X, Chen A 2018 *Chemosphere* **209** 182–190.
[12] Banerjee S, Dionysiou D D, Pillai SC 2015 *Appl Catal B Environ* **176** 396–428.
[13] Najafi V, Ahmadi E, Ziaee F, Omidian H, Sedaghat H 2019 *J Polym Environ* **27** (4) 784–793.
[14] Saeed M, Shakoor A, Ahmad E 2013 *Polym Sci Ser A* **55** (9) 549–555.
[15] Kuwabata S, Takahashi N, Hirao S, Yoneyama H 1993 *Chem Mater* **5** (4) 437–441.
[16] Gangopadyay R. Amitabha De 2000 *Chem Mater* **12** 608.
[17] Song X L, Yan C Y, Huang S T, Zhang M W, Geng B Y, Meng R Bin 2013 *Advanced Materials Research* **750** 1098–1103.
[18] Zagorny M, Bykov I, Melnyk A, Lobunets T, Zhygotsky A, Pozniy A, et al 2014 *J Chem Chem Eng* **8** (2) 118.
[19] Parveen A, Roy AS 2013 *Adv Mater Lett* **4** (9) 696–701.
[20] Yang C, Dong W, Cui G, Zhao Y, Shi X, Xia X, et al 2017 *Acta* **247** 486–495.
[21] Rahman KH, Kar AK 2020 *Mater Sci Semicond Process* **105** 104745.
[22] Mostafaei A, Zolriasatein A 2012 *Prog Nat Sci Mater Int* **22**(4) 273–280.
[23] Kandiel T A, Robben L, Alkaim A, Bahnemann D 2013 *Photochem Photobiol Sci* **12** (4) 602–609.
[24] Hanawalt J D, Rinn H W, Frevel L K 1938 *Ind Eng Chem Anal Ed* **10** (9) 457–512.
[25] Selvan S T, Mani A, Athinarayanasamy K, Phani KLN, Pitchumani S 1995 *Mater Res Bull* **30**(6) 699–705.
[26] Mani A, Athinarayanasamy K, Kamaraj P, Selvan S T, Ravichandran S, Phani KLNN, et al 1995 *J Mater Sci Lett* **14** (22) 1594–1596.
[27] Basavaraja C, Pierson R, Venkataraman A, Basavaraja S 2009 *Macromol Res* **17**(8) 609–615.
[28] Buerger MJ, Azaroff LV 1958 *New York: McGraw-Hill* **238** 342.
[29] Rahman KH, Kar AK 2020 *J Environ Chem Eng* **8** (5) 104181.
[30] Dobrucka R 2017 *Iran J Pharm Res IJPR* **16**(2) 756.
[31] Asha, Goyal SL, Kumar D, Kumar S, Kishore N 2014 *Indian J Pure Appl Phys* **52** (5) 341–347.
[32] Shi L, Wang X, Lu L, Yang X, Wu X 2009 *Synth Met* **159** (23–24) 2525–2529.
[33] Kao S-Y, Lin Y-S, Chin K, Hu C-W, Leung M, Ho K-C 2014 *Sol energy Mater Sol cells* **125** 261–267.
[34] Zhang Q, Bao N, Wang X, Hu X, Miao X, Chaker M, et al 2016 *Sci Rep* **6** 38066.
[35] Seif A, Nikfarjam A, ghassem HH 2019 *Sensors Actuators B Chem* **298** 126906.
[36] Quillard S, Louarn G, Lefrant S, MacDiarmid AG 1994 *Phys Rev B*, **50** (17), 12496.
[37] Socrates G 2004 *John Wiley & Sons*.
[38] Zhang L, Liu P, Su Z 2006 *Polym Degrad Stab* **91** (9) 2213–2219.
[39] Min S, Wang F, Han Y 2007 *J Mater Sci* **42** (24) 9966–9972.
[40] Allen JE, Ray B, Khan MR, Yager KG, Alam MA, Black CT 2012 *Appl Phys Lett* **101** (6) 63105.
[41] Olad A, Behboudi S, Entezami AA 2012 *Bull Mater Sci* **35** (5) 801–809.
[42] Li X, Wang D, Luo Q, An J, Wang Y, Cheng G 2008 *J Chem Technol Biotechnol Int Res Process Environ Clean Technol* **83** (11) 1558–1564.
[43] Wang H, Han B, Lu J, Wu P, Cui W 2020 *Mater Lett* **260** 126906.
[44] Mir FA 2010 *J Optoelectron Biomed Mater* **2** (2) 79–84.
[45] Xia H, Wang Q 2002 *Chem Mater* **14** (5) 2158–2165.
[46] Su S-J, Kuramoto N 2000 *Synth Met* **114** (2) 147–153.

[47] Cionti C, Pina C Della, Meroni D, Falletta E, Ardizzone S 2020 *Nanomaterials* **10** (3) 10030441.
[48] Takpire SR, Waghuley SA 2017 *J Energy Inst* **90** (1) 44–50.
[49] Pankove JI 1975 *Courier Corporation.*
[50] Ahmadi M, Ghasemi MR, Rafsanjani HH 2011 *J Mater Sci Eng* **5** (1) 87.
[51] M. Reda S, M. Al-Ghannam S 2012 *Adv Mater Phys Chem* **02** (02) 75–81.
[52] Kulkarni M, Patil-Sen Y, Junkar I, Kulkarni C V, Lorenzetti M, Iglič A 2015 *Colloids Surfaces B Biointerfaces* **129** 47–53.
[53] Carré A 2007 *J Adhes Sci Technol* **21** (10) 961–981.
[54] Siddiqa AJ, Chaudhury K, Adhikari B 2015 *Res Rev J Med Chem* **1** 43–54.
[55] Leroy F, Müller-Plathe F 2010 *J Chem Phys* **133** (4) 44110.
[56] Psarski M, Pawlak D, Grobelny J, Celichowski G 2019 *Appl Surf Sci* **479** 489–498.
[57] Gongadze E, Velikonja A, Perutkova Š, Kramar P, Maček-Lebar A, Kralj-Iglič V, et al 2014 *Electrochim Acta* **126** 42–60.
[58] Good RJ 1992 *J Adhes Sci Technol* **6** (12) 1269–1302.
[59] Pereira VR, A M Isloor, A K Zulhairun, M N Subramaniam, W J Lau and A F Ismail 2016 *RSC Adv* **6** (102) 99764–99773.
[60] Jlassi K, Radwan AB, Sadasivuni KK, Mrlik M, Abdullah AM, Chehimi MM, et al 2018 *Sci Rep* **8** (1) 1–13.
[61] Deshpande NG, Gudage YG, Sharma R, Vyas JC, Kim JB, Lee YP 2009 *Sensors Actuators B Chem* **138** (1) 76–84.
[62] Arora R, Mandal UK, Sharma P, Srivastav A 2014 *Procedia Mater Sci* **6** 238–243.
[63] Su B, Min S, She S, Tong Y, Bai J 2007 *Front Chem China* **2** (2) 123–126.
[64] Zagorny M, Bykov I 2014 *J Chem Eng Chem Res* **1** (1) 6–14.
[65] Pud AA, Ogurtsov NA, Noskov Y V, Mikhaylov SD, Piryatinski YP, Bliznyuk VN 2019 *Semicond Physics Quantum Electron & Optoelectron* **22** (4) 470-478.
[66] Ogurtsov NA, Noskov Y V, Fatyeyeva KY, Ilyin VG, Dudarenko G V, Pud AA 2013 *J Phys Chem B* **117** (17) 5306–5314.
[67] Li X, Chen W, Bian C, He J, Xu N, Xue G 2003 *Appl Surf Sci* **217** (1–4) 16–22.
[68] Diantoro M, Suprayogi T, Taufiq A, Fuad A, Mufti N 2019 *Mater Today Proc* **17** 1197–1209.
[69] Geethalakshmi D, Muthukumarasamy N, Balasundaraprabhu R 2016 *Mater Sci Semicond Process* **51** 71–80.
[70] Nemade K, Dudhe P, Tekade P 2018 *Solid State Sci* **83** 99–106.
[71] Elumalai NK, Uddin A 2016 *Energy Environ Sci* **9** (2) 391–410.
[72] Babaei Z, Rezaei B, Pisheh MK, Afshar-Taromi F 2020 *Mater Chem Phys* **248** 122879.
[73] Ibrahim M, Bassil M, Demirci UB, Khoury T, Moussa GEH, El Tahchi M, et al 2012 *Mater Chem Phys* **133** (2–3) 1040–1049.
[74] Liu Z, Zhou J, Xue H, Shen L, Zang H, Chen W 2006 *Synth Met* **156** (9–10) 721–723.
[75] Stojanović GM, Radetić MM, Šaponjić Z V, Radoičić MB, Radovanović MR, Popović Ž V, et al 2020 *Appl Sci* **10** (12) 4392.

Optical and Thermophysical Characterization of Fe_3O_4 nanoparticle

Horia.F [1], Khaled Easawi [1], Reda Khalil[2], Said Abdallah [1], Mabrouk El-Mansy[2], and Sohair Negm[1]

[1] Department of Mathematical and Physical Engineering, Faculty of Engineering (Shoubra), Benha University, Cairo, Egypt

[2] Physics Department, Faculty of Science, Benha University

*horya.omer@feng.bu.edu.eg

Abstract

The size dependence of optical and, thermophysical properties of Fe_3O_4 nanoparticle (NPs) has been investigated. Three samples of Fe_3O_4 NPs with different sizes were prepared by the Co-precipitation method and labeled a, b and c. The UV- vis spectra shifted toward the lower energy with increasing the particles size. The energy bandgaps of the synthesized Fe_3O_4 NPs are 1.84 ev,1,73 ev, and 1,47 ev for a, b and c respectively. The sizes of the nanoparticles were estimated using the effective mass approximation (EMA) model, and are 10.7 nm, 20.56 nm, and 30.43 nm. These values are comparable to those found by transmission electron microscope (TEM) and X-ray diffraction (XRD). The Photoacoustic (PA) technique was used to research Fe_3O_4 NPs for thermal diffusivity (α), thermal effusivity (e), and thermal conductivity (k). The determined values of α, e, and k increase as the size of nanoparticles rises and decreases from the bulk by about 33%.

Keywords
Photoacoustic spectroscopy, Magnetite, The thermal conductivity, Thermal diffusivity

1. *Introduction.*

Metal oxides are commonly studied due to their unique electrical and chemical stability, thermal optical, magnetic and rheological, catalytic properties [1]. Ferrites have widespread applications in electronics [2], magneto-optics [3], magnetocaloric refrigeration [4], dynamic sealing, high-density data storage [5], oscillation damping, and catalysis [6–8]. From nanocrystalline ferrites, Fe_3O_4 is the ideal candidate for various biomedical applications because of its cell and biomolecular dimensions [9], low toxicity [10], biocompatibility [11], high saturation magnetization, and easy to control with low magnetic fields [12]. Fe_3o_4 Magnetic particles have physical and chemical properties which could be applied in different fields by transforming them into particles of nanometer scale. These nanoparticles may be used as drug delivery (DDS), magnetic resonance imaging (MRI), and cancer therapy products. The nature of nanoparticles' particle size, magnetic properties, and surface properties was very important for it to be applied to the field [13]. Thermal conductivity was a physical property of a material which shows the ease with substances which could achieve heat energy transmission with molecular motion thermal conductivity of material depending on the substance's chemical composition, the solid 's crystalline structure, the temperature and the pressure under which it was imposed [14]. Massive attention has therefore been paid to investigating its thermal properties to boost the efficiency of solar cell devices and pave the way for new applications as well. [15]. Since the mid-1970, the use of photoacoustic spectroscopy (PAS) to obtain absorption spectrums of solid and fluid substances has attracted considerable interest. The primary source of PA signal results from a periodic heat flow from the solid to the surrounding gas. These can be explained when the beam of light falls on to the sample, the heat produced by the light absorbed in the sample (created by means of non-radiative transition) will diffuse from the sample to the gas through the sample- gas interference. By modulating the light beam, the expansions of the gas layer near the sample create a sound wave transferred from the gas to the microphone [16]. Therefore, the PA signal contains information about the optical absorption within the sample in addition to the way with which the heat is diffused through the sample. The Rosencwaig and Gersho (RG) theory of the PA effect [17] shows that the pressure variations depend on the optical and thermal parameters of the sample. This study aims to synthesize Magnetite Fe_3O_4 NPs of various sizes using the process of co-precipitation. The size-dependent UV-vis spectra have been investigated. The energy band gap of Fe_3O_4 NPs of different sizes was measured. The sizes of nanoparticles have been estimated using effective mass approximation (EMA) [18] and verified by XRD and TEM. Furthermore, thermal diffusivity (α), thermal effusivity (e), and thermal conductivity (k) of different sizes of Fe_3O_4 NPs have been measured using the non-destructive photoacoustic (PA) technique which attracts great interest because it makes them suitable for many technological applications.

2. Experimental work.

2.1Materials

The materials used in the synthesis of Fe_3O_4 NPs are; ferric chloride hexahydrate ($FeCl_3{\cdot}6H_2O$), ferrous chloride tetra-hydrate ($FeCl_2{\cdot}4H_2O$),). Ammonium solution NH_4oH (26% of ammonia) from Analar (England). acetone from Molar Hungary), and nitrogen gas (N_2) with 99.9% purity. Deionized water was used for all experiments from India.

2.2. Synthesis of Fe_3O_4NPs

Co-precipitation is the most widely and simplest method for synthesis magnetics nanoparticles of controlled sizes by precipitate ferrous and ferric salts. Fe_3O_4 were prepared under nitrogen (N_2) gases bubbling through the solution to protect Fe_3o_4 against critical oxidation. 20.25 g ($FeCl_3{\cdot}6H_2O$), and 10.35 g of ($FeCl_2{\cdot}4H_2O$) were dissolved into 250 mL of distilled water. The solution stirring for 80 minutes, chemical precipitation was achieved at 90C under stirring by adding ammonium solution drop by drop at a constant rate until pH ranging between 8 and 14 Completed precipitation of Fe_3O_4 expected [19]. The system was cooled to room temperature, the precipitates were separated by a magnet and washed with distilled water until pH neutral. Finally, Fe_3O_4 washed with acetone and dried in an oven at 70-80C for 1.5 h to prepare the nanoparticles of Fe_3O_4 in powder form, the size and shape depend on the pH value of the solution. We get three different sizes labeled a, b and c. at PH 8-10, 11-12, and 13-14 respectively.

3-*Characterizations*

The absorption bands of the prepared samples have been measured by the UV-670 Jasco double-beam spectrophotometer utilizes a unique, single monochromator design covering a wavelength range from 190 to 2700 nm. The crystal structure of Fe_3O_4 NPs was characterized by X-ray Diffraction (XRD) (Shim ADZU diffractometer type XRD 6000. X-ray diffraction (XRD) is a well-known technique for the characterization of bulk and thin film. Transmission Electron Microscope (TEM) is used for characterizing the size of nanoparticles. TEM (JEOL JEM-2100 operated at 200KV with high-resolution Gatan CCD bottom camera, Orius SC200).

A highly sensitive (PA) technique has been utilized to investigate the thermal parameters (thermal diffusivity (α), thermal effusivity (e), and thermal conductivity (k) of Fe_3o_4 NPs of different sizes. The (PA) technique depends on the detection of pressure fluctuations in a sealed closed-cell created by the periodic thermal waves from the sample. The experimental set-up of the PA technique is shown schematically in figure (1).

The light beam from Stability 2017 Argon Laser was mechanically modulated by an optical chopper (SR540) and focused onto the sample which was mounted carefully inside a PA cell (MTEC Model 300). The sound wave generated from the sample were subsequently detected as an acoustic signal by a highly sensitive electrical microphone fixed in the PA cell. The PA signal was then amplified by a low noise preamplifier and further processed using a lock-in amplifier (Stanford Research System, Model SR830 DSP [20-21]. The PA signal was recorded for varied frequency for each sample.

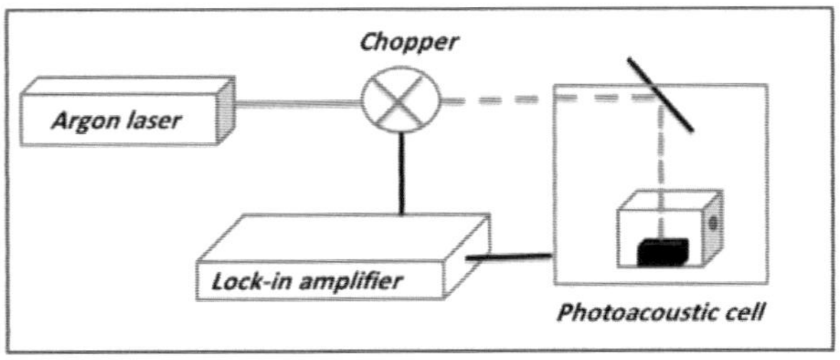

Figure 1. Photoacoustic Technique

4. Results and Discussion.

4.1 *TEM analysis*

TEM analysis has been used to determine the average particle size of Fe_3O_4 NPs The TEM micrographs of samples a,b and c are shown in Fig. 2a, 2b, and 2c respectively. The prepared nanoparticles' images show a nearly spherical morphology. The estimated particle size of a, b, and c samples are 10 nm, 20 nm, and 30 nm respectively.

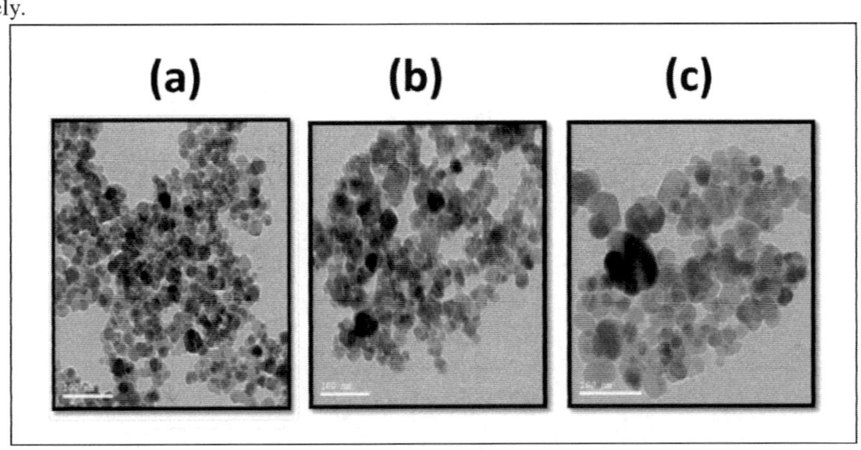

Figure 2. TEM images of the synthesized Fe_3O_4 NPs (a), (b), and(c)

4.2. *UV–Vis absorption*

Optical absorption spectra of samples show a broad absorbance (no sharp absorbance peaks) for all different sizes fig (3), which is mainly due to the half-metallic behavior of Fe_3O_4NPs [22]. The visible absorption is mainly because of the mixture of the ligand near-infrared absorption is assigned to the inter valence field transition and charge-transfer transition, while the charge-transfer transition between the metal ions. The absorption edges are 372 nm,383 nm, and 404 nm for the samples a, b, and c respectively

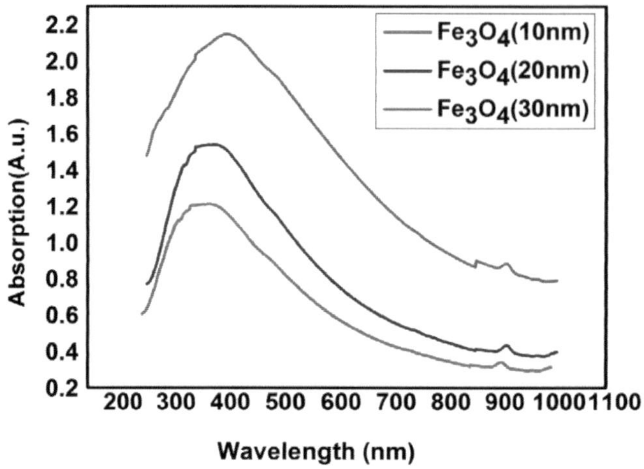

Figure 3. the absorption spectra of the prepared Fe3o4with different sizes 10,20and30nm respectively

The electronic transitions are the essential parameter of the optical absorption. These transitions are controlled by certain selection rules, which can be expressed by using Tauc's equation [23].

$$\alpha h\upsilon = A(h\upsilon - E_g)^n \qquad (1)$$

where $h\nu$ is the energy of the incident photons, α is the absorption coefficient, A is a constant, h is Planck's constant, and n is a constant with allowed values of 1/2 and 2 and not allowed values of 3/2 and 3 for direct and indirect transitions correspondingly. The values of n were detected in the range (0.925-1.083) which predicts the electronic allowed direct transition. The values of band-gap are determined by the Tauch-plot curve of $\alpha h\upsilon^2$ value on hυ. The interception of the extrapolation of this linear part to the energy axis gives energy bandgap E_g Figure 4 shows the difference of $(\alpha h\nu)^2$ with photon energy hv for the three samples (a,b, and c). The values of E_g for Fe_3o_4 NPs are 1.84, 1,73, and 1,47ev.

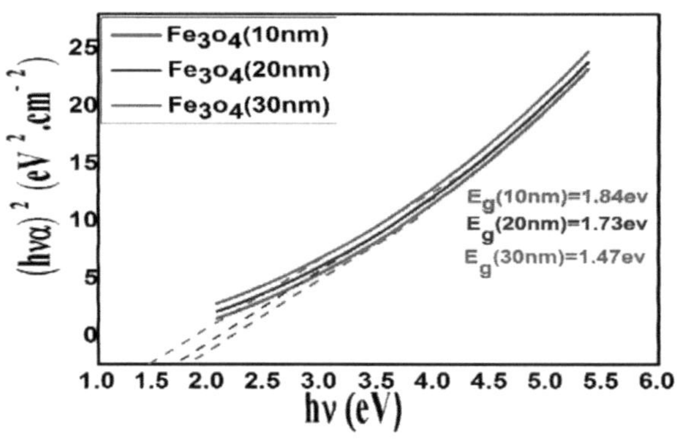

Figure 4. $\alpha h\nu^2$ was plotted versus hv for synthesized samples Fe_3O_4 NPs10, 20 and 30nm respectively

Using the calculated bandgap and the "Effective Mass Approximation" model (*EMA*) the size of Fe_3O_4 NPS is calculated. The EMA model is given by [24].

$$E_{gn} = E_{gb} + \frac{h^2}{8R^2}\left[\frac{1}{m_e} + \frac{1}{m_h}\right] - \frac{1.8e^2}{4\pi\varepsilon\varepsilon_0 R} \quad (2)$$

where Egn is the electronic bandgap for bulk Fe_3o_4 (0.1 ev), R is the average radius of NPs, m_e is the electron effective mass ($100m_o$), m_h is the hole effective mass ($100m_o$), $m_0 = 9.11*10^{-31}$ Kg and ε is the dielectric constant forFe_3O_4[24]. The calculated particles sizes are 10.3 nm, 20 .2 nm, and 30.3 nm which are comparable to those obtained by TEM measurements.

4.3 X-Ray diffraction analysis.

XRD measurements were used to recognize the crystalline structure of the products. Fig (5) showed that the XRD peaks can match fine with the characteristic peaks of the inverse cubic spinel (reflection peak at $2\theta=35.6^{\circ}$) structure (JCPDS 19-629), which shows the crystalline structure of Fe_3O_4 NPs.

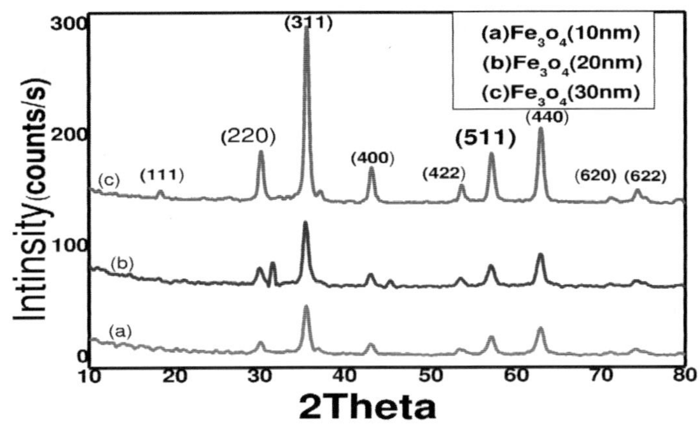

Figure 5. the XRD spectra of synthesized Fe_3o_4 with different sizes(a), (b)and(c) respectively.

pattern could be readily identified as the pure cubic phase of Fe_3O_4. The average crystalline size calculated using the Debye–Scherrer equation [25].

$$D = \frac{0\cdot 9\lambda}{B\ cos\theta} \quad (3)$$

Where β is the full-width at half-maximum (FWHM) value of XRD diffraction lines, the wavelength λ= 0.15405 nm and θ is the half diffraction angle of 2θ [26-28]. The value 2θ is the average of 30.1°,35. °6,43.2°,53.5°, and 56.95 at the diffraction peaks for each sample. The average crystalline size of Fe_3O_4 NPs is 10.7 nm, 20.20 nm, and 30.43nm in close agreement with TEM and optical measurements.

4.4 Thermal measurements

The thermal properties of Fe_3O_4 nanoparticles were studied with the Photoacoustic (PA) technique. The samples in powder form were compressed into disks (under a hydraulic pressure 30 Ton/inch) with the thickness (0.62 mm). For each appropriate (depth profile analysis) the PA signal amplitude was reported at different chopping frequencies (f). Fig 6 shows the plots of ln PA signal amplitude versus lnf for synthesized samples (a), (b), and (c) respectively. A distinct slope transition is observed at a frequency f_c (the sample transitions from thermally thick to thermally thin) where the crossover occurs. The thermal diffusivity (α), for all samples, was then calculated using the relation [29].

IOP Conf. Series: Materials Science and Engineering **956** (2020) 012016 doi:10.1088/1757-899X/956/1/012016

$$\alpha = f_c\, L^2 \quad m^2/sec \qquad (4)$$

where L is the thickness of the sample. The measured values of α are shown in Table 1. It easily observes that as the size of Fe_3o_4 NPs increase the thermal diffusivity increase.

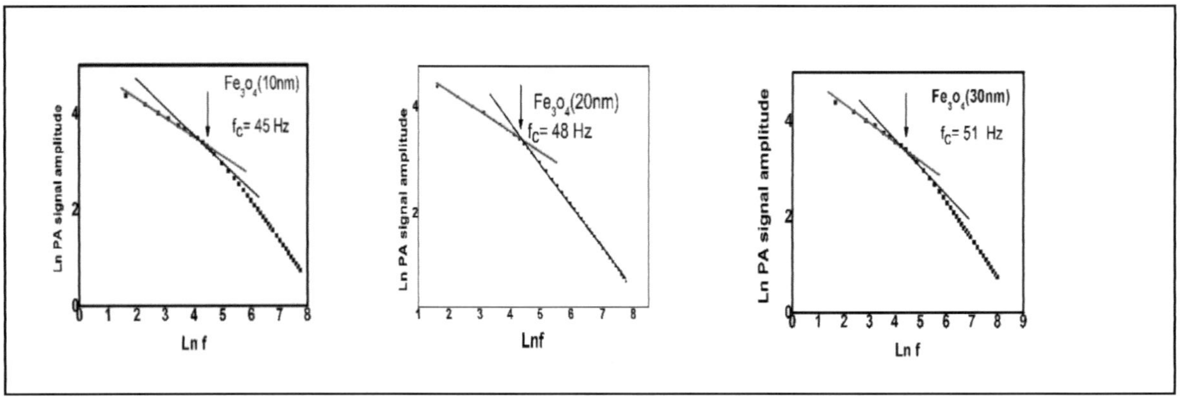

Figure 6. ln PA signal amplitude versus lnf for synthesized Fe_3O_4 10,20&30nm respectively

The thermal effusivity (e) of the samples was also calculated using the PA technique, where the PA signal amplitude q is given for optically opaque and thermally thick samples by [30- 32]

$$q = \frac{B}{e}\frac{1}{f} \qquad (5)$$

where $B = \dfrac{I_0\ \gamma P_0 \alpha_g^{1/2}}{2\ 2\pi l_g T_0}$,(I_0) is the incident light strength, (Y) is the specific heat ratio, (P_0) is the ambient pressure, (α_g) is the gas thermal diffusivity, (l_g) is the length of the gas column, (T_0) is the ambient temperature and (f) is the modulation frequency. Using Si as a reference sample, with 15670 $Ws^{1/2}m^{-2}K^{-1}$ thermal effusivity [29]. The Si reference sample is used to extract the B constant.

IOP Conf. Series: Materials Science and Engineering **956** (2020) 012016 doi:10.1088/1757-899X/956/1/012016

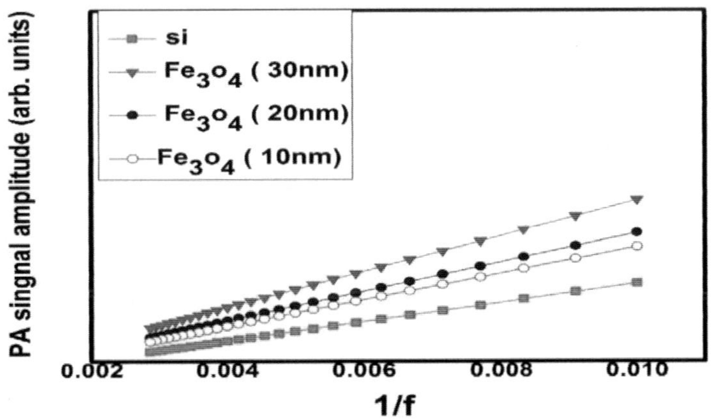

Figure 7. the variation of PA signal amplitude with inverse frequency for the reference sample and prepared NPs.

Fig (7) show the relationship between the PA signal amplitude and the inverse of the chopping frequency for reference sample (si) and three different sizes of Fe_3o_4, the slopes were obtained by linear fitting for the relation between (1/chopping frequency) and PA amplitude. The resultant thermal effusivities are given in table 1. The equivalent values of thermal conductivity $K=e\sqrt{\alpha}$ are calculated and given in table (1)

Table1. The measured values of the three Parameters for different sizes of Fe_3o_4 NPs

Sample Fe_3O_4	Thermal diffusivity (α) (10^{-5} m^2/s)	Thermal effusivity (e) (10^3 $Ws^{1/2}$ m^{-2} K^{-1})	Thermal conductivity (k) (W/m.K)
10nm	1.72	1.18	4.9
20nm	1.84	1.28	5.5
30nm	1.96	1.35	6

The values of (α.e and k) are increasing as the size of nanoparticles increases and decreases from its bulk value. The value of (k) for sample c is about 66 % from bulk value (9.7W/mK) [33]. Such decreases in (k) of Fe_3O_4 NPs from the bulk value are agreement with N. Rajkumar et al [33]. In their work, they obtained a decrease in (k) about 59 % for particles size 10 nm from the bulk value. In such a study, they explained the decrease of k of Fe_3O_4 NPs from the bulk value based on phonon. When the particles are small, phonon confinement increases, and thus the collision rate increases (mean free path decreases) resulting in reduced thermal conductivity. Moreover, As the size of the nanomaterial decrease, the surface-to-volume ratio increases, and the presence of increased surfaces due to roughness additionally provides the phonon with a scattering mechanism. This additional scattering decreases the mean free path of the phonon which causes the resistance in the heat flow resulting in a reduction of the thermal diffusivity and conductivity [34 - 35].

Conclusion

Magnetite nanoparticles synthesized by chemical co-precipitation method. The prepared Fe_3O_4(NPs) nanoparticles were characterized by TEM, UV-vis, and XRD. The average particle sizes estimated by TEM images are 10 nm,20 nm, and 30 nm. The UV-vis absorption spectra of Fe_3O_4 shifted toward the visible region and the redshift of broad an absorbance spectrum to longer wavelength with increasing particle size. The particle sizes are confirmed by an effective mass approximation model and XRD data. Furthermore, the Photoacoustic (PA) technique has been used to study the thermal diffusivity (α), thermal effusivity (e), and thermal conductivity (k) of Fe_3O_4 NPs. The values of (α.e and k) are increasing as the size of nanoparticles increases and decreases from its bulk value. The value of (k) for sample c is about 66 % of the bulk value it decreases with a decrease in particle size. The decrease in thermal conductivity as the particle size decreased is due to the increase in phonon containment and thus the increase in collision rate (mean decrease in the free path) leading to a decrease in thermal conductivity.

References

[1] Kim, J, K., Lee, J, H., & Lee, S, H. (2004). Journal of Magnetism and Magnetic Materials, 279(2-3), 173-177.

[2] Zargar, M., Hamid A, A., Bakar F, A., Shamsudin, M, N., Shameli, K., Jahanshiri, F., & Farahani, F. (2011). Molecules journal,16(8),6667–6676.

[3] Makovec, D., Kosak, A., &Drofenik, M. (2004). International Journal of Nanotechnology,15, S160-S166.

[4] McMichael, R, D., Shull, R, D., Swartzendruber, L, J., Bennett, L, H., &Watson, R, E. (1992). Journal of Magnetism and Magnetic Materials,111(1-2), 29-33.

[5] Shen, L., Laibinis, P, E., &Hatton, T, A. (1999). Langmuir,15, 447.

[6] Chou, S, Y., Wei, M, S., Krauss, P, R., & Fischer, P, B. (1994). Journal of Applied Physics, 76(3), 6673-6675.

[7] Raj, K., & Moskowitz, R. (1990). Journal of Magnetism and Magnetic Materials, 85(1-3), 233-245.

[8] Wang, X., Peng, Q., & Li, Y. (2005). Crystal Growth and Design Journal, **5**(2), 391-393.

[9] Ferreira, H, A., Graham, D, L., Freitas, P, P., & Cabral, J, M, S. (2003). Journal of Applied Physics,93(10), 7281-7286.

[10] Zhang, L., Dong, W, F., &Sun, H, B. (2013). Nanoscale journal,5(17),7664–7684.

[11] Jain, T, K., Morales, M, A., Sahoo, S, K., Leslie- Pelecky, D, L., & Labha-salwar, V. (2005). Molecular Pharmaceutics,2(3), 194-205.

[12] Kim, D, K., Mikhaylova, M., Wang, F, H., Kehr, J., Bjelke, B., Zhang, Y., Tsakalakos, T., & Muhammed, M. (2003). Chemistry of materials. Mater, 15(23), 4343-4351.

[13] Masala, O., & Seshadri, R. (2005). Chemical physics letters, 402(1-3), 160-164.

[14] Gope, P, C., Singh, V, K., & Rao, D, K. (2015). Journal of Reinforced Plastics and Composites,34(13): 1075-1089.

[15] Badawi, A., Al-Hosiny, N., Abdallah, S., Negm, S., & Talaat, H. (2012). Journal of Materials Science and Engineering, 2 (1) 1-6.

[16] Mandelis, A., Teng, Y, c., & Royce, b, S. (1979). Journal of Applied Physics, 50 (11), 7138-7146.

[17] Rosencwaig, A., & gersho. (1876). Journal of Applied Physics, 47,64.

[18] Sathyamoorthy, R., Sudhaga, r P., Saravana, R., & Sathy, T, M. (2010), Journal of Advanced Research, 46, 99 – 103.

[19] Sophie, L., Delphine, F., Marc, P., Alain, R., Caroline, R., Luce, V, E., N. & Robert, M. (2008). Chemical Reviews, 108, 2064-2110.

[20] okasha, A, M., T, Abdallah., Mohamed, M, B., Anis, B, B., Negm, S., & Talaat, H.(2010).journal of physics. conference series,214,19-23.

[21] Abdallah ,S., Abdallah, T., Mohamed ,M ,B., & Negm ,S .(2008). The European physical journal , 8,365-368.

[22] Hsia, C, H., Chen, T, Y., & Song, H. (2008). Nano Letters journal,8(2), 571-576.

[23] Ahmadi, M., Ghasemi, M, R., & Rafsanjani, H, H. (2011). Journal of Materials Science and Engineering,5,87-93.

[24] Lahore, P., & Jain, p. (2015). International journal Research Chemistry and Environment,4,38-43.

[25] El Ghandoor, H., Zidan, H, M., Mostafa, M, H., & Ismail, M, I. (2012). International Journal of Electrochemical Science, 7, 5734 – 5745.

[26] Victoria, L., Calero, C., & Carlos, R. (2007). Journal of Magnetism and Magnetic Materials,314, 60.

[27] Zhang, Y., Yang, Z., Liu, Y., ChunLong, F., &Yan, G. (2010). Journal of Magnetism and Magnetic Materials,322(22), 3470-3475.

[28] Kumar, V., Rana, A., Yadav, M, S., &Panta, R, P. (2008). Journal of Magnetism and Magnetic Materials, 320(11), 1729-1734.

[29] Almond, D, P., & Patel, P,.M (1996). champman& Hall series in accounting,13,978-8808

[30] Abdallah, S., (2004). Egyptian journal of solids,27,451.

[31] Awad, H., Abdallah, T., Mohamed, M, B., Easawi, K., Negm, S., & Talaat, H.(2010).journal of physics: conference series,214(1),012130.

[32] Abdallah, S., easwi, k., Negm, S., Youssef, G, M., El-brolossy, T, A., & H, talaat. (2003). Review of scientific insturments,74,848.

[33] Rajkumar, N., Umamahaeswari, D., & Ramachandran, K. (2010). International Journal of Nanoscience, 3,243–250.

[34] Arora, N., Deepika, P, J., & Pachauri, U. (2018). Materials Chemistry and Physics,217,235-241.

[35] Arabshahi, H. (2011). International Journal of the Physical Sciences, 6(13), 3155–3161

IOP Conf. Series: Materials Science and Engineering **956** (2020) 012017 doi:10.1088/1757-899X/956/1/012017

Effect of local additive (BM2010) on high performance concrete under sulphate attack

A. Megahid Ahmed[1], Omar A. Fargal[1], and Mostafa Abd Elrazek [2], Ashraf Abd Eltawab[3]

[1] Department of Civil Engineering, Faculty of Engineering, Asyut University, Asyut, Egypt.
[2] Department of Civil Engineering, High Institute of Engineering &Technology, EL-MINYA, Egypt.
[3] New Asyut City, New Urban Communities Authority, Egypt.

E-mail: dr.mostafa.abd.k@mhiet.edu.eg, dr.mostafa.abd.k@gmail.com.

Abstract. Many of reinforced concrete structures especially underground and undersea are subjected to aggressive environmental conditions throughout their exploitation. These conditions are represented by the aggressive sodium sulphate attack from surrounding soil or underground water. When based materials of cement are exposed to sodium sulphate attack produce gypsum and ettringite by its chemical reactions. Formation of gypsum plays a very important role within the damage of materials. According to that the actual compressive strength of concrete elements decreases accompanied with large deformations of concrete. Consequently, a higher reduction of the construction durability occurs, so using of effective and economic admixture for concrete elements protection against sulphate attack or the other environmental conditions is needed.

The main purpose of this research is to study through an experiment the effect of the local admixture (BM2010) which contained in its composition wastes from petroleum industries, silica fume, Wastes of coke factory and its mechanism effect on the compressive strength of concrete specimens subjected to external sulphate attack.

The experimental results showed that Compressive strength of concrete specimens modified with the organic admixtures (BM2010) at age of 128 days hardened in salt solution with concentration (9%NaCl + 12% Na_2SO_4) increased by about 15% compared to the control specimen with (Addicrete dm2) for sulphate resistance cement (S.R.C) & 3.6% compared to the control specimen (without admixture) for ordinary Portland cement (O.P.C). The absorption of concrete specimens modified with the organic admixture (BM2010) at age of 128 days hardened in salt solution with concentration (9%NaCl + 12% Na_2SO_4) is less than concrete specimens modified with (Addicrete dm2) by about 29%percent in case of (S.R.C) but in case of (O.P.C) it is more by about 16% percent. The experimental results showed that, the optimum composition of all components of the suggested admixture (BM 2010) was successfully and experimentally achieved.

1. Introduction

Marine concrete occupies a vital half and becomes one in all the most issues. sulfate attack of concrete in marine setting has been wide investigated. The degradation behaviours and harm phenomena of concrete served in marine setting are very difficult. it's the interaction between sulfate ions, calcium hydroxide which ends up in cracks and degradation of concrete materials. historically, the employment of pozzolanic materials, made of silica and alumina, as partial replacement of cement is one in all economical ways to rising sulfate attack resistance (1).

Successful usage of supplementary cementitious materials (SCMs) to replace cement helps to solve relevant environmental issues such as CO_2 emission and waste management as well as saving energy. It is also well known that the inclusion of SCMs in concrete mixtures enhances durability and improves mechanical properties. However, when only taking environmental and economic pressure into consideration, the high replacement levels of most of the SCMs usually lead to losses in strength at early ages (1).

Lee et al., (2) used Silica fume, Nie et al.,(3) used pozzolanic reaction of fly ash, Atahan et al. (4) used ground granulated blast furnace slag and Chalee et al., (5) used rice husk ash in sulphate attack resistance. Metakaolin have been proved to be beneficial in enhancing the mechanical properties and sulphate attack of conventional concrete (6).

When concrete structures are in a sulphate environment, external sulphate ions will enter the concrete, which generates a diffusion of sulphate ions in concrete, better known as an external sulphate attack. Two causes of this attack are: -

- sulphate ions diffuse from a high concentration to a low concentration in concrete pores.
- sulphate ions react with the hydration products of cement.

The concrete sulphate attack is caused by the cracking and spalling of concrete due to the expansion of the sulphate attack products, resulting in a concrete strength loss. Sulphate ions from outside environment will attach the concrete through two patterns: diffusion and chemical reaction. However, it is still difficult to quantify the sulphate attack product and its interaction in concrete (7).

Concrete subjected to Sulphate attack is one of concrete durability problem in concrete which used in industrial application and concrete exposed to sea water or marshy organic soil where sulphate concentration (SO4) in water or soil is enough to initiate sulphate attack in concrete, which pose in turn of Expansive Ettringite compound within the hardened cement paste & resulting disruption of concrete start from concrete core to surface due to its continuous increase in volume within the same specified volume. Considering the durability problem of concrete exposed to sulphate exposure there are many ways of preventive measures are usually adopting in practice either internally or externally (8).

Sodium Sulphate Na_2SO_4 attack in concrete is more serious and hindering than Calcium sulphate attack in concrete due to high solubility of Sodium sulphate than Calcium sulphate. The sodium sulphate is reacting with hydration product of cement Calcium hydroxide present in concrete to form Calcium sulphate (gypsum), which is then react with Tri calcium aluminate (C_3A) to form Calcium Sulpho-aluminates (Ettringite) which is hydrophilic in nature & absorb water resulting increase in volume within the hardened concrete & disruption of concrete will start. The step wise reaction mechanism of Sodium sulphate in concrete are hereby stated below in equation (1) and (2).

Step-I: $Na_2SO_4 + Ca(OH)_2 = CaSO_4 .2H_2O + NaOH$ (1)

Step-II: $3CaSO_4 + 2H_2O + 3C_3A + 26\ H_2O = C_3A + 3CaSO_4 + 32H_2O$ (Ettringite) (2) (8).

Industrial solid wastes are inducing severe environmental problems, but the problem can be overcome by reusing them as construction materials. concrete incorporating solid waste materials can not only promote the recycling of solid waste, but also provide high sulphate attack resistance (9).

Admixtures have long been recognized as an important component of concrete used to improve its performance. The original use of admixtures in cementitious mixtures is not well documented. It is known that cement mixed with organic matter was applied as a surface coat for water resistance or tinting purposes. It would be a logical step to use such materials, which imparted desired qualities to the surface, as integral parts of the mixture. The use of natural admixtures in concrete was a logical progression. Chemical admixtures confer certain beneficial effects on concrete, including reduced water requirement, increased workabi1ity, controlled setting and hardening, improved strength, and better durability.

Many approaches have been adopted to investigate the role of chemical admixtures. One approach is to determine the state of the admixture in concrete at different times of curing. The admixture may remain in a free state as a solid or in solution or interact at the surface or chemically combine with the constituents of cement (10).

A.Razek (11) found that the suggested organic admixture (BM 2010) has a good effect on the properties of the concrete mixes which prepared and produced from petroleum industries wastes, silica fume and naphthalene sulfonate having main Chemical component as follows. (Tested by Analytical Chemistry Unit – ACAL Assiut University)

Parameter	Description
Chloride	0.0132 % (wt/vol)
Sulphate	0.1416 % (wt/vol)
Calcium oxide	0.1624 % (wt/vol)
Aluminum oxide	0.0560 % (wt/vol)

Concrete incorporates a low resistance to chemical attack. There are many chemical agents that react with concrete; however, 2 types of attacks are most common, namely, natural action and sulphate attack. Sulphate attack on concrete end up in the conversion of hydrous merchandise of cement to ettringite, gypsum and alternative phases, and to the destabilization of the primary strength provided calcium silicate hydrate (C-S-H) gel. The formation of ettringite and gypsum is common in cementitious materials exposed to most types of sulphate solutions. The enlargement ensuing from sulphate attack is generally attributed to the formation of these two compounds, although there is some controversy surrounding the exact mechanisms causing expansion. Chemical agents essentially react with certain compounds of the hardened cement paste and the resistance of concrete to chemical attack therefore depends largely on the type of cement used. The resistance of concrete to chemical attack has improved with increase of its permeability (12).

When cement-based materials square measure exposed to sulfate attack, gypsum and ettringite square measure made by chemical reactions of sulfate and $Ca(OH)_2$, C_3A. Formation of mineral plays a very important role within the injury of materials. there's a detailed relationship between $Ca(OH)_2$ content and gypsum formation. Ettringite formation ends up in cracking and growth of the material. growth is said to the water absorption of crystalline ettringite. So, it's necessary to extend the resistance of concrete against salt attack (12).

Increasing the concrete alkali content by adding alkali addition to the mixture water has harmful effects on most mechanical properties (compressive, splitting, direct tensile, and flexure strengths) of concrete. Admixtures, that contain in their compositions organic materials with restricted doses have several benefits such as: concrete compressive strength and concrete tensile strength, concrete durability subjected to aggressive environmental conditions, appropriate resistance to corrosion and really low value (12)

Many papers studied the effect of organic admixtures in concrete duo to their positive influence on the physical and mechanical properties of R.C. elements not only on their early time of hardening, but also for the period of their exploitation in the building site. Investigations of victimization chemical admixtures for the development of concrete properties in Egypt show that, they are not sufficient, and the component of those admixtures area unit foreign by the origin and not made in Egypt. Therefore, the system of our constructions utterly depends on the foreign corporations. So, these components become deficient and costly. These area unit the particular issues for business and system of Egyptian industries. it's renowned that the organic materials haven't the solubility with water. So, during this analysis the solubility of those parts with water was achieved by victimization of calcium oxide and superplasticizers. Investigations regarding victimization complicated organic

IOP Conf. Series: Materials Science and Engineering **956** (2020) 012017 doi:10.1088/1757-899X/956/1/012017

admixtures, that contain in their composition alkali wastes from oil industries for the development of cement materials properties were dole out. (12).

From the accessible literature few studies on the mechanism effect of organic admixtures, containing wastes from petroleum industries, on the resistance of concrete elements against sulphate and chloride attack. The aim of this study is to analyse the effect of the local additive (BM2010) on the strength, sulphate and chloride resistance of high-performance concrete.

2. Experimental

The materials used for the research work are 2 forms of cement (ordinary Portland cement (O.P.C) , sulphate resistance cement (S.R.C)).The coarse aggregate used for the experiment is of crushed Basalt rock, Fine aggregate of coarse sand and the drinking water is used. The admixture used are local admixtures (BM 2010), (Addicrete dm2).

Tested specimen

Experimental tests have been carried out on (54) standard concrete cubes specimens (15x15x15 cm). All specimens tested under static loading up to failure. This study mentioned the effect of suggested admixture (BM 2010) by the following parameters.

- Additive type
- Cement type

These specimens were as follow:
This group divided into 2 groups:

- Group (1): O.P.C (concrete with ordinary Portland cement (Almohandes cement N.42.5))
- Group (2): S.R.C (concrete with sulphate resistance cement (Altameer cement N.42.5))

Each of these 2 groups is divided as following:

- 9 cubes without additives
- 9 cubes with additive (BM2010)
- 9 cubes with additive (Addicrete dm2)

Group (1) plus Group (2) are (54 cubes). All cubes were cured in sulphate solution (9% NaCl + 12% Na_2SO_4) for 100 days after 28 days were cured in drinking water.
Concentration of surrounded sulphate & chloride solutions (Na_2SO_4+ NaCl): Specimens of group were cast and hardened in freshwater conditions until 28-days. After that, were immerged and hardened in concentration of sodium sulphate and sodium chloride solution for a period of 100 days as given in Table (1). Concentration of salt solution. 9%NaCl + 12%Na_2SO_4. All cubes were tested under axial static compression load

Table (1): Concentrations of salt solution

Salts	Concentrations by weight of water
NaCl	9%
Na_2SO_4	12%

Test Procedure

(54 cubes) were cured in fresh water until of 28 days after that put in concentration of sulphate & chloride solutions for a period of 100 days. The testing machine (Control 1500 KN) was used. Each specimen was loaded axially and gradually keeping the rate of loading constant. The concrete specimens were tested under static axial compression loading after 28 & 128 days hardening in fresh water or in salt solutions.

IOP Conf. Series: Materials Science and Engineering **956** (2020) 012017 doi:10.1088/1757-899X/956/1/012017

Materials

Cement

1- Ordinary Portland cement (almohandes cement)

Table (2): Physical cement prop.

Test		Result	ESS No. 2421-1/2005
Cement expansion (Le Chatelier)		1 mm	< 10 mm
Specific gravity		3.15	-
Setting time	Initial	165 minute	≥ 45 minutes

Table (3): Chemical cement prop.

Test	Result	ESS No. 4756 – 1 /2009
Loss on ignition	1.0 -2.0%	≤ 5%
Insoluble residue	0.3 % - 0.6%	≤ 5%
% of sulphate ions content (So3)	2.5 % - 3.0%	≤ 3.5%
% of chloride ions content (Cl)	0.02 – 0.04	≤ 0.1%

Table (4): Cement compressive strength

Test		Result	ESS
Compressive strength	2 days	20.7 N/mm^2	-
	28 days	42.7 N/mm^2	42.5 N/mm^2

2- Sulphate resistance cement (altameer cement)

Table (5): Physical cement prop.

Test		Result	ESS No. 2421-1/2005
Cement expansion (Le Chatelier)		1.2	< 10 mm
Specific gravity		3.15	-
Setting time	Initial	113 minute	≥ 45 minutes

Table (6): Chemical cement prop.

Test	Result	ESS No. 5325/2006
Loss on ignition	2.64	≤ 5%
Insoluble residue	.39	≤ 5%
% of sulphate ions content (So3)	2.5	≤ 3.5%
% of chloride ions content (Cl)	.04	≤ 0.1%

Table (7): Cement compressive strength

Test		Result	ESS
Compressive strength	2 days	18.7 N/mm2	-
	28 days	42.9 N/mm2	≥ 42.5 N/mm^2

IOP Conf. Series: Materials Science and Engineering **956** (2020) 012017 doi:10.1088/1757-899X/956/1/012017

Water

Drinking water was used for mixing concrete

Aggregate

The used coarse aggregate was basalt and local natural sand was used as fine aggregate in experimental work. The properties of the used aggregate are given in table (8)

Concrete mixes

Concrete mixes were used to produce high performance concrete having 28-days cubic compressive strength 600 kg/cm^2.
Concrete mixes components shown in Table (8).

Table (8): Details and properties of concrete mixes (Sulphate resistance cement)

Mix No.	Cement (Kg/m^3)	Sand (Kg/m^3)	Coarse Agg. (Kg/m^3)	Additivs		Water (L/m^3)	W/C
				Dose % of cement	Type		
1	450	663	1162	2.0 1.5	G NN-1 +Fss	162	0.36
2	450	663	1162	1.6 1.1 0.9	G NN-1 +Fss + BM	162	0.36
3	450	663	1162	1.6 1.1 0.4	G NN-1 +Fss + Addic	140	0.31

Slump=15-20cm

Table (9): Details and properties of concrete mixes (Ordinary Portland cement) :

Mix No.	Cement (Kg/m^3)	Sand (Kg/m^3)	Coarse Agg. (Kg/m^3)	Additives		Water (L/m^3)	W/C
				Dose % of cement	Type		
4	450	663	1162	2.0 1.7	G NN-1 +Fss	148 L	0.33
5	450	663	1162	2.2 1.1 0.9	G NN-1 +Fss + BM	165 L	0.37
6	450	663	1162	2.2 1.1 0.4	G NN-1 +Fss + WP	169 L	0.38

Slump=15-20cm

Table (10): Type & percent of admixtures

Type of admixture	Optimum dose (% of cement wt.)
Control	0
Addicrete dm2(control)	0.4
BM2010(suggested)	0.9

IOP Conf. Series: Materials Science and Engineering **956** (2020) 012017 doi:10.1088/1757-899X/956/1/012017

Table (11): Physical, Mechanical and Chemical properties of used aggregate

Property	Gravel	ESS	Sand	ESS
Volume weight (t/m^3)	1.72	-	1.62	-
Specific gravity	2.63	-	2.5	-
% of clay & Fine (dust) by weight	0.85%	1.0%	2.8	≤ 3%
Crushing value %	15.76	≤ 30	-	-
% of chloride ions content (Cl)	0.018	≤ 0.04	0.06	≤ 0.06
% of sulphate ions content (So_3)	0.035	≤ 0.4%	0.04	≤ 0.04
F.M	6.2	-	2.42	-

Table (12): Results of sieve analysis of used aggregate

Sieve size, mm	% Passing by weight	
	Gravel	Sand
37.5 mm	100	-
20 mm	98.94	-
10 mm	74.58	-
5 mm	5.59	98.55
2.36 mm	1.48	94.09
1.18 mm	0.45	86.13
0.60 mm	-	65.68
0.30 mm	-	11.77
0.15 mm	-	1.66

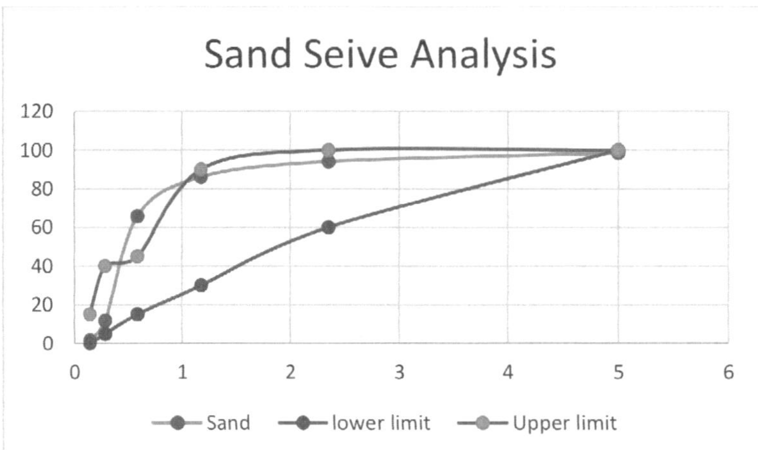

Fig. (1): % Passing of Sand sieve analysis

IOP Conf. Series: Materials Science and Engineering **956** (2020) 012017 doi:10.1088/1757-899X/956/1/012017

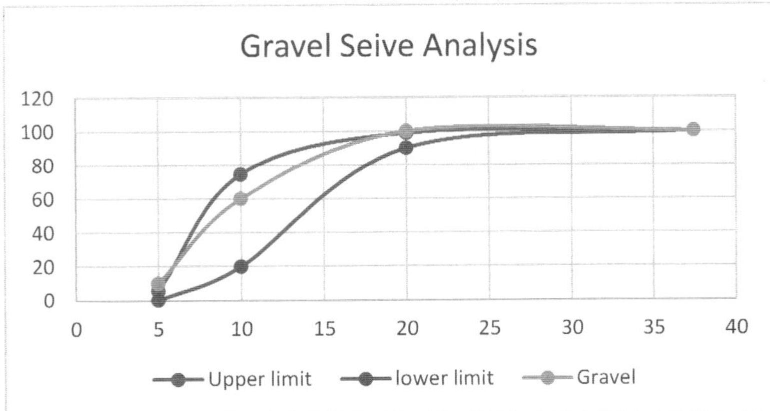

Fig. (2): % Passing of Gravel sieve analysis

Different types of admixtures:

1- Suggested organic admixture (type BM 2010).
2- Control admixture (commercial admixture) (type addicrete dm2)
3- Control sample (without admixture)

Suggested admixture (Type BM 2010) Which contained in its composition wastes from petroleum industries, silica fume, wastes of coke factory and pure water reduces permeability and produces considerably higher resistance to salt and attacking chemicals (Optimum dose 0.9% of cement weight). A.Razek (2014)

Commercial admixture (Addicrete dm2): is a chemical waterproofing admixture for concrete and mortar.

Description: It reduces permeability and produces considerably higher resistance to salt and attacking chemicals. Optimum dose 0.4% of cement weight. (From Company Technical data sheet).

Sedrete Gnn1: high range water reducing and super-plasticizer for concretes.

Description: is a high range water reducing and super-plasticizer and increases compressive strength and sitting time.

Dosage: 0.6-3.5 liter for each 100 kg/m cement according to the required stress

(From Company Technical data sheet).

Sedrete Fss: high range water reducing and super-plasticizer for concretes.

Description: is used as super-plasticizer that can reduce water, improve workability, and increase early and ultimate stresses.

Dosage: 0.6-4 % litter for each 100kgm cement according to the required stress. (From Company Technical data sheet).

3. Results and discussion

The main variables studied in this research on concrete mixes and hardened specimens were:- Effect of Two different types of organic admixtures, Suggested (BM2010), commercial admixture (Addicrete dm2) and control (without admixture) which cured on salt solution with concentration of (9%NaCl + 12% Na_2SO_4) for a period of 128 days. Two types of cements were used (S.R.C), (O.P.C) & high-performance concrete. Fifty four cubes were used in this research (2 types of cement×3 types of admixtures×9cubes/admixture).Test results of all cubic concrete specimens which hardened in aggressive environmental condition are presented in Tables (11 , 12) , fig. (1,2).

Effect of optimum dose of each admixture on compressive strength and absorption of concrete specimens cured in salt solution with concentration of (9%Nacl, 12% Na_2SO_4) for (S.R.C) & (O.P.C) in table (11), (12).

Table (13) (S.R.C)

Additive type	Fc28	% increase	Fc128	% increase	Absorption
Without	633	0	713	0	0.97
BM2010	641	1.3%	752	5.5%	0.97
Addicrete dm2	614	-3%	589	17.4%	1.25

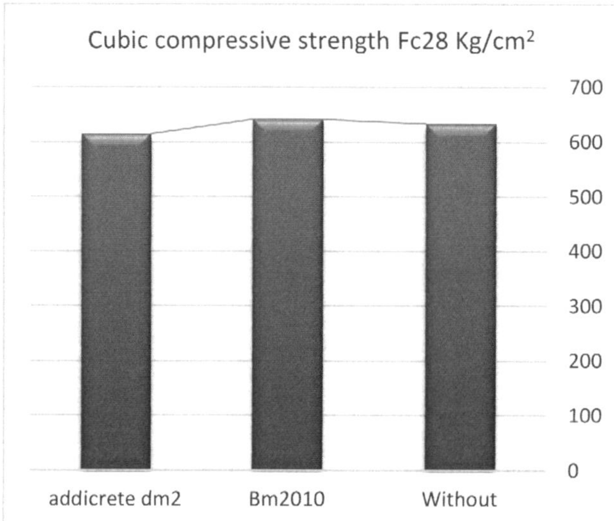

Fig. (3): Cubic compressive strength of concrete specimens versus age of concrete (fc28) for (S.R.C) with salt solution (9%Nacl+12% Na_2SO_4)

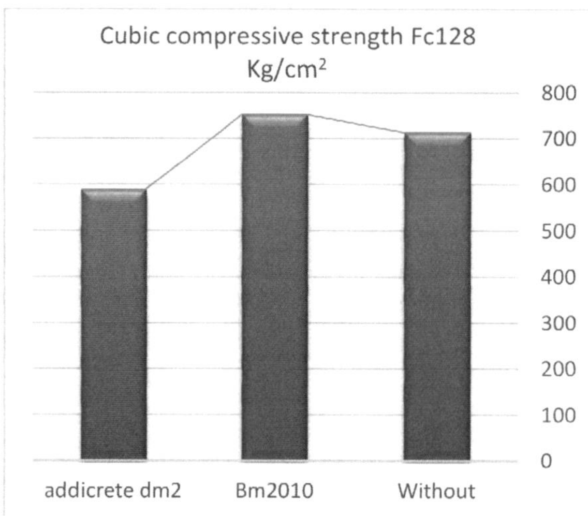

Fig. (4): Cubic compressive strength of concrete specimens versus age of concrete(fc128) for (S.R.C) with salt solution (9%Nacl+12% Na_2SO_4)

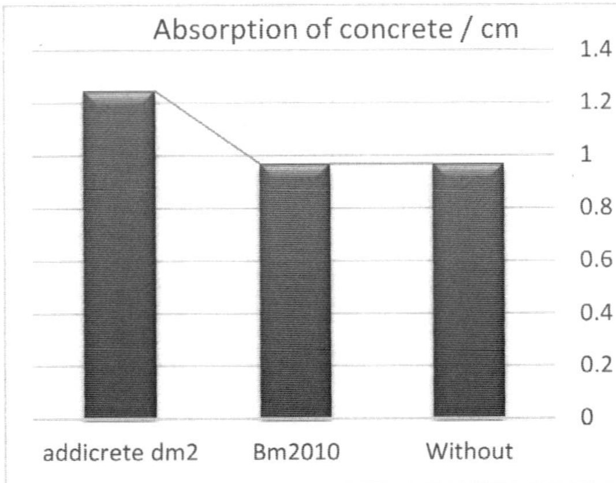

Fig. (5): Absorption of concrete specimens after 128 days age cured in (9%Nacl, 12% Na_2SO_4) for (S.R.C)

Table (14) (O.P.C)

Additive type	Fc28	% increase	Fc128	% increase	Abs.
Without	442	0	467	0	0.88
BM2010	553	25.1%	596	27.6%	0.92
Addicrete dm2	642	45.2%	603	39.1%	0.77

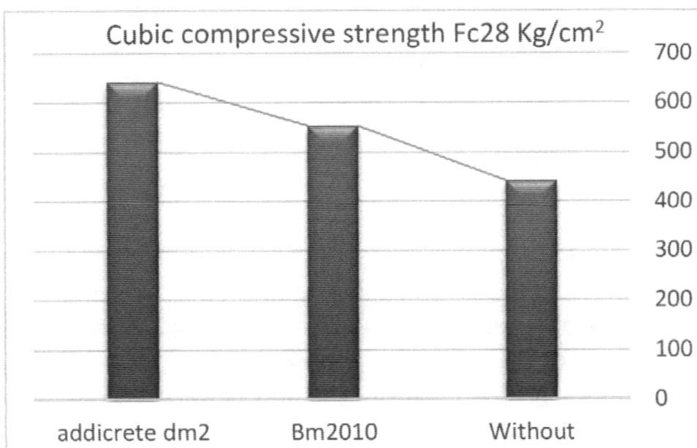

Fig. (6): Cubic compressive strength of concrete specimens versus age of concrete (fc28) for (O.P.C) with salt solution (9% Nacl + 12% Na_2SO_4)

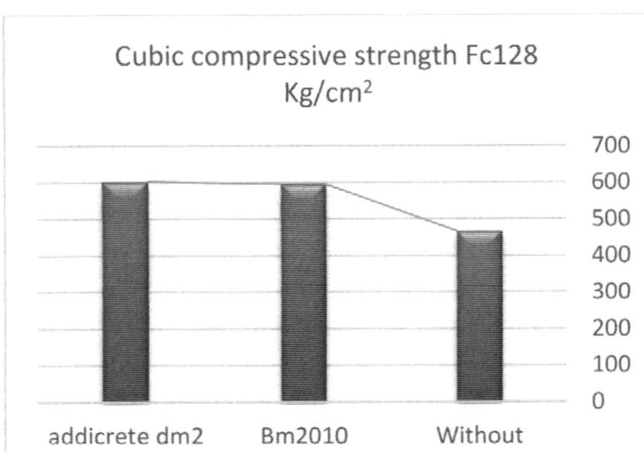

Fig. (7): Cubic compressive strength of concrete specimens versus age of concrete(fc128) for (O.P.C) with salt solution (9%Nacl+12% Na_2SO_4)

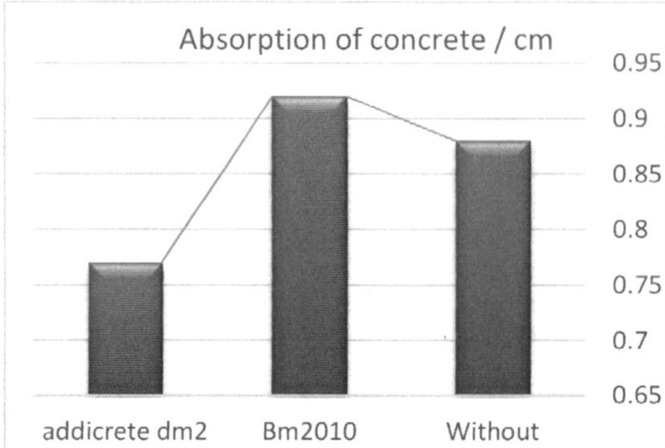

Fig. (8): Absorption of concrete specimens after 128 days age cured in (9% Nacl, 12% Na_2SO_4) for (O.P.C)

Compressive Strength of Concrete Specimens Exposed to Salt Solutions:

Compressive strength of all types of concrete containing the admixtures and cured in salt solution with concentration of (9%NaCl + 12% Na_2SO_4) are shown in Tables (11 , 12) , fig. (3,4,6,7).

The suggested organic admixture (BM 2010) has a good effect on increasing compressive strength of concrete cured in aggressive condition (9%NaCl + 12% Na_2SO_4) either in (S.R.C) or (O.P.C) cement type and helped the concrete specimens successfully to reach to high performance concrete. Tables (11 , 12) , fig. (3,4,6,7).

- Compared with control specimens (Addicrete dm2) the organic admixture (BM 2010) increase the compressive strength by about 4.4% & 28% at 28 &128 days respectively for specimens immersed in salt solution with concentration (9%NaCl + 12% Na_2SO_4) in case of use (S.R.C) (Tables 11, fig.3,4).
- Compared with control specimens(without admixture) the organic admixture (BM 2010) increase the compressive strength by about 1.3% & 5.5% at 28 &128 days respectively for specimens immersed in salt solution with concentration (9%NaCl + 12% Na_2SO_4) in case of use (S.R.C) (Tables 11, fig.3,4).
- Compared with control specimen (Addicrete dm2) The organic admixture (BM 2010) decrease the compressive strength by about 14% & 1% at 28 & 128 days respectively for specimens immersed in salt solution with concentration of (9%NaCl + 12% Na_2SO_4)in case of use (O.P.C) (Tables 12, fig. 6,7).
- Compared with control specimen (without admixture) The organic admixture (BM 2010) increase the compressive strength by about 25% & 28 % at 28 & 128 days respectively for specimens immersed in salt solution with concentration of (9% NaCl + 12% Na_2SO_4) in case of use (O.P.C) (Tables 12, fig. 6,7).
- The organic admixture(BM 2010)increase the Concrete compressive strength for specimens of(S.R.C) after 28,128 days by about 16&26% respectively more than (O.P.C) for specimens immersed in salt solution with concentration of (9% NaCl + 12% Na_2SO_4). (Tables 11 , 12 , fig.3,4, 6,7).
- The organic admixture (BM 2010) decrease the absorption for specimens of (O.P.C) after 128 days by about 5% less than (S.R.C) for specimens immersed in salt solution with concentration of (9%NaCl + 12% Na_2SO_4). This meaning the organic admixture (BM 2010) increase the permeability resistance of water in case of (O.P.C) more than (S.R.C) in like this aggressive condition (Tables 11 , 12 , fig. 5,8).
- Compared with control specimens(Addicrete dm2) & control specimens (without admixture) the organic admixture (BM 2010) decrease the absorption by about 29% after 128 days in case of (Addicrete dm2) but in case of control specimens (without admixture) it is~equal (this is for

specimens immersed in salt solution with concentration (9%NaCl + 12% Na_2SO_4)) in case of use (S.R.C) (Tables 11 fig.5).

- Compared with control specimens (Addicrete Wp) & control specimens (without admixture) the organic admixture (BM 2010) increase the absorption by about 16% after 128 days in case (Addicrete dm2) but in case of control specimens(without admixture)it is increase by about 4% (this is for specimens immersed in salt solution with concentration (9%NaCl + 12% Na_2SO_4)) in case of use(O.P.C) (Tables 12 fig.8)
- The organic admixture (BM 2010) decrease the absorption for specimens of (O.P.C) after 128 days by about 5% less than (S.R.C) for specimens immersed in salt solution with concentration of (9% NaCl + 12% Na_2SO_4). This meaning the organic admixture (BM 2010) increase the permeability resistance of water in case of (O.P.C) more than (S.R.C) in this aggressive condition (Tables 11 , 12 , fig. 5,8).

4. Conclusion

The results of the experiments carried out on the concrete mixes and specimens with different additives (Suggested additive (BM2010), control additive (Addicrete dm2) and control (without additive) and cured in salt solution with concentration of (9%NaCl + 12% Na_2SO_4) for a period of 128 days, the following conclusions can be drawn out:

1- Compressive strength of concrete specimens modified with the organic admixtures (BM2010) at age of 128 days cured in (9% NaCl + 12% Na_2SO_4) increases by about 28% compared to the control specimen (Addicrete dm2) for (S.R.C) type.

2- Concrete specimens compressive strength modified with the organic admixtures (BM2010) at age of 128 days cured in (9% NaCl + 12% Na_2SO_4) increases by about 5.5% compared to the control specimen (without admixture) for (S.R.C) type.

3- Concrete specimens compressive strength modified with the organic admixtures (BM2010) at age of 128 days cured in (9% NaCl + 12% Na_2SO_4) increases by about 1% compared to the control specimen (Addicrete dm2) for (O.P.C) type.

4- Concrete specimens Compressive strength modified with the organic admixtures (BM2010) at age of 128 days cured in (9% NaCl+12% Na_2SO_4) increases by about 28% compared to the control specimen (without admixture) for (O.P.C) type.

5- The absorption of concrete specimens modified with the organic admixtures (BM2010) at age of 128 days cured in (9% NaCl + 12% Na_2SO_4) decreases by about 29% compared to the control specimen (Addicrete dm2) for (S.R.C) type.

6- The absorption of concrete specimens modified with the organic admixtures (BM2010) at age of 128 days cured in (9% NaCl + 12% Na_2SO_4) equals compared to the control specimen (without admixture) for (S.R.C) type.

7- The absorption of concrete specimens modified with the organic admixtures (BM2010) at age of 128 days cured in (9% NaCl + 12% Na_2SO_4) increases by about 16% compared to the control specimen (Addicrete dm2) for (O.P.C) type.

8- The absorption of concrete specimens modified with the organic admixtures (BM2010) at age of 128 days cured in (9% NaCl + 12% Na_2SO_4) increases by about 4% compared to the control specimen (without admixture) for (O.P.C) type.

9- (BM2010) is better than (Addicrete dm2) in case of using (S.R.C) for fc28 and fc128 and absorption.

References

[1] Guo, L., Wu, Y., Duan, P., & Zhang, Z. (2020). Improving sulphate attack resistance of concrete by using calcined Mg-Al-CO3 LDHs: Adsorption behavior and mechanism. Construction and

Building Materials, 232, 117256.
[2] Lee, S. T., Moon, H. Y., & Swamy, R. N. (2005). Sulphate attack and role of silica fume in resisting strength loss. Cement and Concrete Composites, 27(1), 65-76.

[3] Nie, Q., Zhou, C., Li, H., Shu, X., Gong, H., & Huang, B. (2015). Numerical simulation of fly ash concrete under sulphate attack. Construction and Building Materials, 84, 261-268.
[4] Atahan, H. N., & Arslan, K. M. (2016). Improved durability of cement mortars exposed to external sulphate attack: The role of nano & micro additives. Sustainable cities and society, 22, 40-48.
[5] Chalee, W., Sasakul, T., Suwanmaneechot, P., & Jaturapitakkul, C. (2013). Utilization of rice husk–bark ash to improve the corrosion resistance of concrete under 5-year exposure in a marine environment. Cement and concrete composites, 37, 47-53.
[6] Ren, D., Yan, C., Duan, P., Zhang, Z., Li, L., & Yan, Z. (2017). Durability performances of wollastonite, tremolite and basalt fiber-reinforced metakaolin geopolymer composites under sulphate and chloride attack. Construction and Building Materials, 134, 56-66.
[7] Min, H., Sui, L., Xing, F., Tian, H., & Zhou, Y. (2019). An effective transport model of sulphate attack in concrete. Construction and Building Materials, 216, 365-378.
[8] Laskar, A. A., & Ghosh, P. (2019). Influence of Coal Tar Epoxy Paint on Concrete Exposed to Sulphate Exposure. International Research Journal of Engineering and Technology (IRJET), 6(9) 2395-0072
[9] Tang, Z., Li, W., Ke, G., Zhou, J. L., & Tam, V. W. (2019). Sulphate attack resistance of sustainable concrete incorporating various industrial solid wastes. Journal of Cleaner Production, 218, 810-822.
[10] Ahmed, A. M., Rashwan, M. M., & Razek, M. A. (2013). Performance of a Local Proposed Economical Additive on the Cement Mortar Properties Affected by Aggressive Environmental Conditions. Journal of Engineering Sciences, Faculty of Engineering, Assiut University, 41(2), 421-440.
[11] Mostafa A.Razek. (2014). Local economical additive for producing high durable concrete for water tanks. PHD. Dissertation Department of Civil Engineering, Assiut University.
[12] Ahmed, A. M., Rashwan, M. M., & Razek, M. A. (2013). MECHANISM EFFECT OF A SUGGESTED LOCAL AND ECONOMICAL ADDITIVE ON THE COMPRESSIVE STRENGTHAND DURABILITY OF CONCRETE. 41 (1), 29-47
[13] Abd Elrazek, M., & Gamal, Y. A. S. (2020). Producing High Workable & High Strength Concrete Mix Having 6 Hour Retardation.

IOP Conf. Series: Materials Science and Engineering **956** (2020) 012018 doi:10.1088/1757-899X/956/1/012018

Plasmonic Yolk-Shell Nano Structures for Solar Cells Absorbance Enhancement

Hadeer H. AbdelAziz [1], Tamer A. Ali [1,2] and Nadia H. Rafat [1]

[1] Faculty of Engineering, Cairo University, Giza, Egypt
[2] Communications and Information Engineering Program, Zewail City of Science and Technology, October Gardens, 6th of October, Egypt

E-mail: hadeer.eldanasoury.j@eng1.cu.edu.eg

Abstract
In this research we study the enhancement in power absorbance of thin film silicon solar cells by plasmonic yolk-shell (YS) nano particles. Plasmonic YS nano particles are simulated on top/bottom position of thin film Si layers. We suggest different structures of YS and simulate the behaviour of such structures using a FDTD simulations. We calculate the limiting photocurrent taking into consideration the enhanced absorbance of the cell. The simulation results show multi-peaked and broad band power absorbance enhancement all over the solar spectrum. The power absorbance enhanced into solar cell by 64% over planar ultra-thin 300 nm Si cell.

1. Introduction

To achieve high absorbance, the first prototype of solar cells was based on thick films. About half of the cost of solar cell comes from material cost [1]. The designs of the second-generation solar cell suggest reducing the amount of materials and facilitating fabrication methods. Thin film solar cells (TFSC) offer cost reduction and large scalability fabrication. However (TFSC) faces the problem of low absorption of light especially for incident photons with low energy [2]. Trapping light techniques are used to boost up the efficiency of (TFSC) [3], [4], [5]. Many researchers studied the effects of plasmonic nano particles on TFSC for Si material [6], [7], [8]. In a phenomenon called localized surface plasmon resonance (LSPR), plasmonic nano structures can squeeze, confine and localize electromagnetic waves (EMW) inside nano range. Free electrons couple with photons of EMW to make collective oscillations "plasmons". This phenomenon integrated with solar cells and the cells are called plasmonic solar cell (PSC). Plasmonics are very sensitive to geometry, size, material and location of nano particles. Top/bottom solar cell's surface located plasmonic nano particles are proposed as an efficient way to increase absorbance of (TFSC) by scattering and creating additional paths inside solar cell [9]. Gold and silver are the dominant used materials in the field of plasmonic in the visible region [10]. Plasmonic yolk-shell (YS) is a special form of core-shell nano particle with a void shell layer. The name inspired from the "rattle toy" as indication of movable core [11]. Nano YS has 2 forms (i) metal core -void -dielectric shell and (ii) metal core -void -metal shell. This second type can be mono or bi-metallic as Au-void-Ag or Au-void-Ag. The void layer could be liquid or air. Plasmonic nano YS attracts many researches because of its unique properties that arisen from its asymmetry, the existence of a void layer, and the ability of core movement. The LSPR of plasmonic asymmetric nano particles has many peaks over simple spherical or symmetric shells structures. As a hybrid structure (composed of 2 shells and a core), plasmonic nano YS properties can be interpreted by (PHT) theory. In [12] the plasmonic Au/Pd nano YS resonance is a hybrid or a mixture of the plasmon resonance of spherical Au core and the Pd nano shell. The narrow LSPR of Au sphere, and the broaden spectrum of Pd nano shell hybridize with each other to form the plasmon resonance of plasmonic nano YS. The experimental calculations agree with theoretical discrete dipole approximations (DDA). In [13] they report the fabrication of Au-void-Silica nano YS by etching of Au core for the use of photo catalytic reaction. In [14] the availability to fabricate Ag -void -silica nano YS was introduced. Plasmonic (YS) are mainly used in solar energy conversion applications and photo catalysis [15],[16],[17] and can easily be fabricated by galvanic replacement which is a chemical reaction to produce the YS from nano shell [18]. The main properties that affects the (YS) structure is the core and shell material, the sizes of core and shell and the void space. By The proper engineering of nano YS, it can boost the efficiency of thin film solar cell. In our study, seeking for improving solar cell absorbance we will study the effects of noble metal plasmonic nano YS on the enhancement of solar cells.

2. Structure and Methods

Nano YS shape has many configurations with many material parameters. In this study we are interested in nano YS shape with core and shell plasmonic materials like gold and silver. The YS shape studied here, as shown in Fig. (1) is a metal core with radius r and two shells, the first shell is void or hollow shell and the second shell is another metal. All the YS are with core radius of r=60 nm, shell 1 of radius R1=70 nm and shell 2 of radius R2=100 nm. We changed the plasmonic materials to see the changing effects in solar cell power absorbance enhancement.

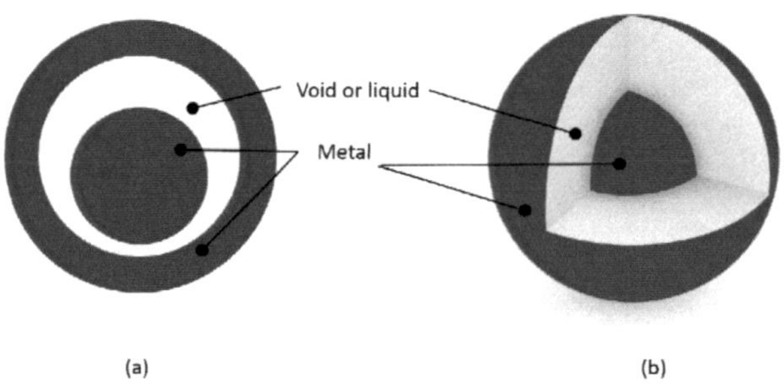

Fig. (1): Schematic diagram of plasmonic YS (a) 2D (b) 3D

Fig. (2) shows the structure of a thin Si solar cell layer of 300 nm with nano yolk-shells on its surface. An array of plasmonic YS nano particles with periodicity of 400 nm is considered. We performed a 3D FDTD study. In our simulation we assume a perfect matched layer (PML) along the z-direction and periodic boundary conditions (PBCs) along x and y-directions. A parametric study is performed for the PSC with changing the materials of core and shells of the nano yolk shell structure. In this simulation, the AM1.5 solar spectrum [19] is represented by a plane wave source injected along the z-direction. To calculate the amount of absorbed power, two power monitors are placed inside cell. The first monitor positioned at the cell surface to measure reflected power; another monitor positioned at the bottom of the cell. Therefore, the absorbed power is the subtraction of both. The quantity power absorption is defined as neither reflected nor transmitted power [20].

$$\boldsymbol{A(\lambda)} = \mathbf{1} - \boldsymbol{R}\,(\boldsymbol{\lambda}) - \boldsymbol{T(\lambda)} \qquad (1)$$

where, $\boldsymbol{A(\lambda)}$ is the absorbance, $\boldsymbol{R(\lambda)}$ is the reflectance and $\boldsymbol{T(\lambda)}$ is the transmittance. To quantify the solar cell absorbance enhancement, we define a Figure of Merit (FoM) as the ratio between the power absorbed of PSC to that of a reference planar cell. FoM is calculated considering that a single electron-hole pair is produced for every photon absorbed with energy bigger than the band gap of Si [21].

$$\boldsymbol{FoM} = \frac{\int \lambda\ P_{PSC}(\lambda)\ I_{AM1.5}\,(\lambda)d\lambda}{\int \lambda\ P_{Planar}\,(\lambda)\,I_{AM1.5}\,(\lambda)d\lambda} \qquad (2)$$

where, P_{PSC} is the power absorption in solar cell with plasmonic YS and P_{planar} is the power absorption in planar cells. Also, FoM can be calculated at every wavelength of the solar spectrum.

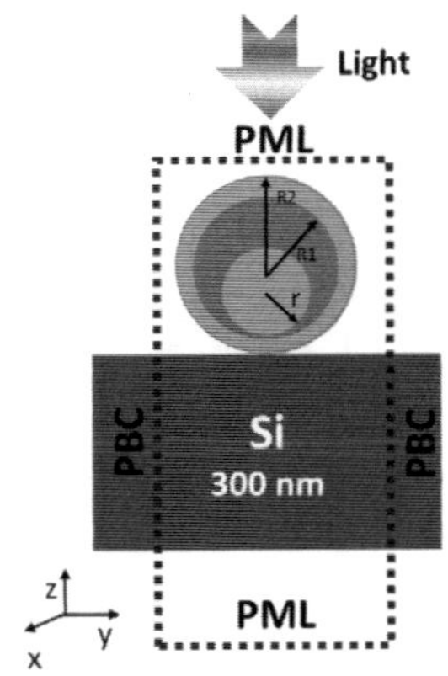

Fig. (2): Schematic of plasmonic solar cell with YS nano particles

3. Results and Discussion

3.1 Plasmonic YS on a Si Substrate

The plasmonic YS Ag-void-Ag and Ag-void-Au are used either on the top surface of Si or at the bottom surface. The absorbance, $\boldsymbol{A(\lambda)}$, of thin film planar 300 nm Si solar cell with different top/bottom Ag-void-Au YS is shown in Fig. (3). The absorbance spectra of thin film Si solar cell enhanced by using bottom located Ag-void-Au YS as it reaches 47% absorbance of the spectra at λ=514 nm. Fig. 4 shows the distribution of the relative power absorption per unit volume (log scale) in W/mμ^3 at different three peaks of the absorbance curve of Fig. (3). Fig. 4 (a), (b) show the distribution of the relative power absorption per unit volume for top YS Ag-void-Au at λ= 514 nm and 950 nm, respectively. Fig. 4 (c) shows the distribution of the relative power absorption per unit volume of the bottom YS at λ=514 nm. These figures show that at λ=514 nm the absorbance enhancement results from the bottom Ag-void-Au YS is higher than that of top located YS. For λ=950 nm, top located Ag-void -Au YS causes higher absorbance inside Si layer. In comparison to the reference cell, Fig. 4.6 (d, e) show that the absorption is enhanced effectively at λ = 514 nm and 950 nm, respectively due to plasmonic YS. The reference cell suffers from high reflectance and transmittance in compare to plasmonic solar cell.

IOP Conf. Series: Materials Science and Engineering **956** (2020) 012018 doi:10.1088/1757-899X/956/1/012018

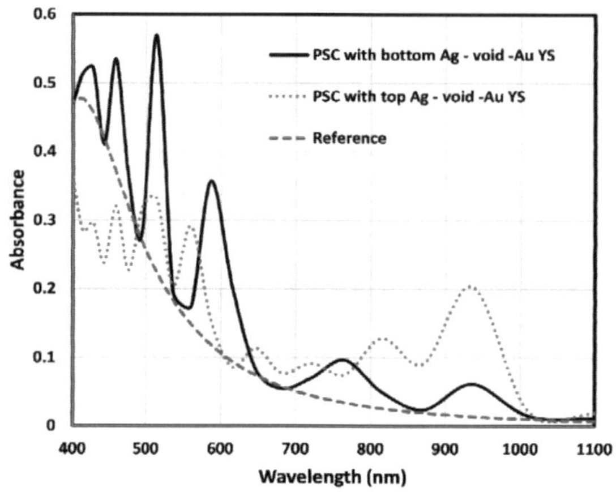

Fig. (3) Absorbance of Si 300 nm with top and bottom Ag-void-Au YS particles (R2=100 nm, R1=70 nm and r=60 nm)

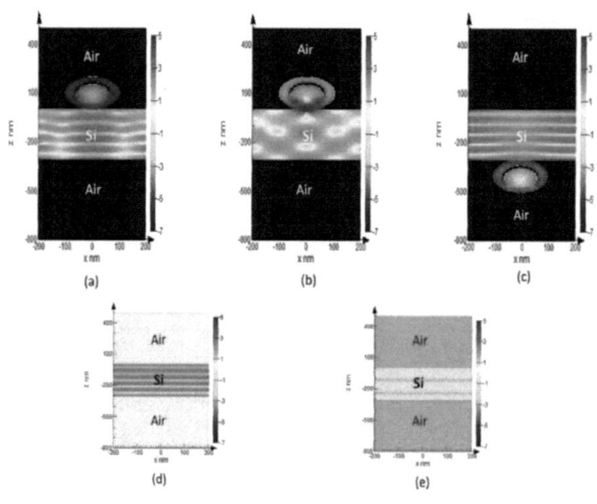

Fig. (4): The distribution of the relative power absorption per unit volume (W/μm3) (log scale) of PSC with Ag −void−Au (R2= 100 nm, R1= 70 nm and r= 60 nm) located at (a) top at λ = 514 nm, (b) top at λ = 950 nm, (c) bottom at λ = 514 nm, and reference 300 nm Si cell at (d) λ = 514 nm, (e) λ = 950 nm

Plasmonic effects of Ag-void-Ag YS in enhancing solar cell absorbance is shown in Fig. (5). The usage of plasmonic top/ bottom located Ag-void-Ag YS enhances the absorbance of ultra-thin 300 nm Si layer solar cell. Bottom located plasmonic Ag-void-Ag YS exceeds 64% absorbance of the solar spectrum at λ=514 nm. The top located Ag-void-Ag YS enhances the absorbance compared to the reference cell, but the enhancement of the bottom located YS is much better. For each corresponding absorbance peak, the distribution of the relative power absorption per unit volume (log scale) in W/mμ^3 is calculated for top Ag-void-Ag plasmonic solar cell at λ= 560 nm and 816 nm in Fig. 6 (a, b). For plasmonic solar cell with bottom Ag-void-Ag, the distribution of the relative power absorption per unit volume (log scale) is calculated at λ= 514 nm and 721 nm in Fig. 6 (c, d). These figures show that the Ag-void-Ag YS enhanced the absorption of light incident on Si solar cell all over the spectrum in comparison to the relative power absorption distribution profile of the planar cell Fig. 6 (e-h) at λ=560, 816, 514 and 721 nm, respectively.

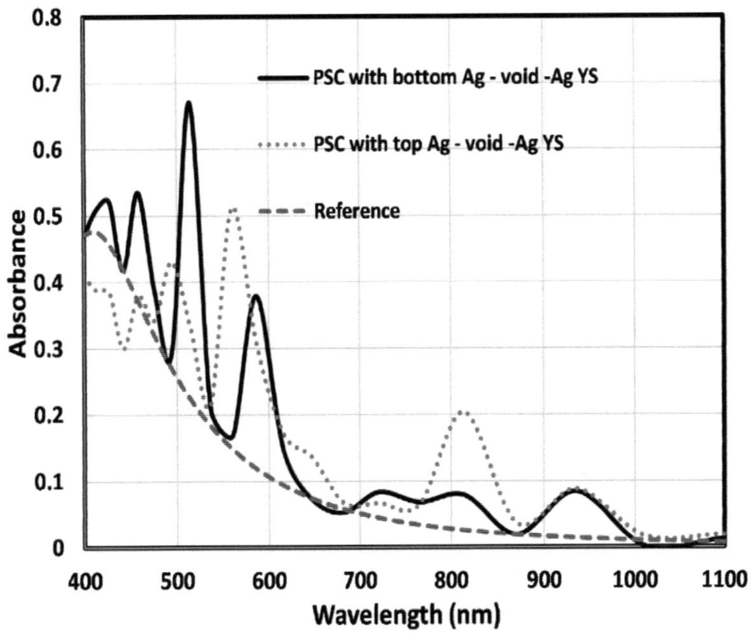

Fig. 5 Absorbance of Si 300 nm with top and bottom located Ag-void-Ag (R2=100 nm, R1=70 nm and r=60 nm).

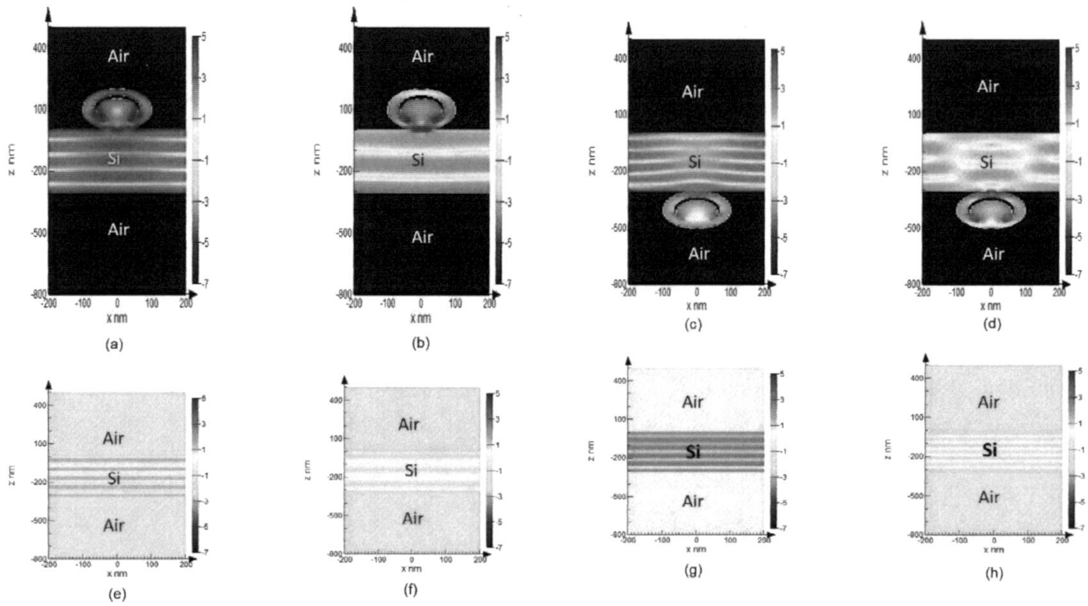

Fig. (6): The distribution of the relative power absorption per unit volume (W/μm^3) (log scale) of PSC with Ag-void-Ag Ag (R2=100 nm, R1=70 nm and r= 60 nm) top located at wavelength (a) 560 nm and (b) 816 nm and bottom at wavelength (c) 514 nm and (d) 721 nm.

Fig. (7) shows the FoM of different PSC with different YS at each wavelength of the solar spectrum. FoM is always greater than one, except at some wavelengths of top Ag-void-Au. the inclusion of plasmonic top/bottom YS nano particles have an enhancement effect on the absorption of Si. Integrating of the bottom Ag-void-Ag YS shows advantage to other YS as the absorbance enhancement has many peaks at λ= 514 nm, 586 nm, 765 nm, 936 nm, and 1100 nm, respectively.

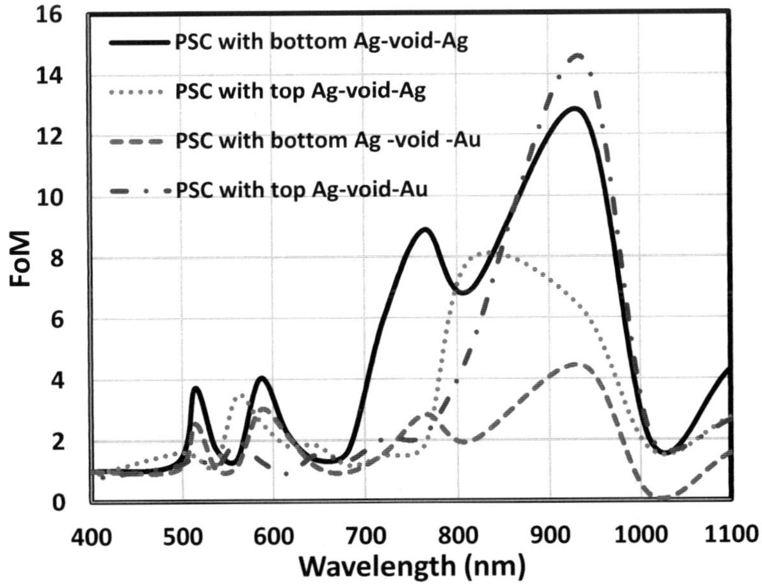

Fig.7 Enhancement of absorbance w.r.t lambda of PSC with different YS nano structures

The calculation of FoM using equation (2) is introduced in Table 1 for a reference cell, PSC with top/bottom Ag-void-Ag and Ag-void-Au with (r=60 nm, R1= 70nm, R2=100 nm). FoM indicates the overall absorbance enhancement all over the spectrum compared to the reference cell. Generally using plasmonic YS nano particles enhances the absorption of ultra-thin 300 nm Si layer. The bottom Ag-void-Ag YS has the highest figure of merit of 1.64. The enhancement in the absorption affects the values of ideal short circuit current of solar cell. The plasmonic YS nano particles could trap light efficiently inside Si especially when placed on bottom surface of Si layers.

Table 1: FoM and limiting short circuit current of different solar cells. Nano particles, if used, are yolk shell nano rattles of radius r=60 nm, R1= 70nm, R2=100 nm and the embedded shell is void

Type of Solar Cell	Nano Particles	Location	FoM
Planar Si 300 nm	—	—	1
PSC	Ag-void-Ag	Top	1.51
PSC	Ag-void-Ag	Bottom	1.64
PSC	Ag-void-Au	Top	1.40
PSC	Ag-void-Au	Bottom	1.47

Conclusion

Our aim is to produce highly absorptive thin film Si solar cells. The plasmonic YS nano particles are placed on top or bottom surface of ultra-thin 300 nm Si layer to explore the plasmonic enhancement effect on the absorption of Si solar cell. Plasmonic Ag-void-Ag YS nano particles increased the absorbance of thin film Si solar cell especially when placed on the bottom surface. Compared to a reference planar cell, the PSC with YS nano particles enhanced the absorption with a figure of merit of 1.64 and with as high-quality absorption in the near infra-red region for $\lambda > 600$ nm where the reference cell absorbance in this region is very small.

References

[1] Mohanty P and Tyagi A 2015 Introduction to Solar Photovoltaic Technology *Food, Energy, and Water* (Elsevier) pp 309–48
[2] Shah A V, Schade H, Vanecek M, Meier J, Vallat-Sauvain E, Wyrsch N, Kroll U, Droz C and Bailat J 2004 Thin-film silicon solar cell technology *Prog. photovoltaics Res. Appl.* **12** 113–42
[3] Ferry V E, Verschuuren M A, Li H B T, Verhagen E, Walters R J, Schropp R E I, Atwater H A and Polman A 2010 Light trapping in ultrathin plasmonic solar cells *Opt. Express* **18** A237--A245
[4] Morawiec S, Mendes M J, Priolo F and Crupi I 2019 Plasmonic nanostructures for light trapping in thin-film solar cells *Mater. Sci. Semicond. Process.* **92** 10–8
[5] Tabrizi A A and Pahlavan A 2020 Efficiency improvement of a silicon-based thin-film solar cell using plasmonic silver nanoparticles and an antireflective layer *Opt. Commun.* **454** 124437
[6] Pillai S, Catchpole K R, Trupke T and Green M A 2007 Surface plasmon enhanced silicon solar cells *J. Appl. Phys.* **101** 093105
[7] Ouyang Z, Zhao X, Varlamov S, Tao Y, Wong J and Pillai S 2011 Nanoparticle-enhanced light trapping in thin-film silicon solar cells *Prog. Photovoltaics Res. Appl.* **19** 917–926
[8] Mandal A and Chaudhuri P 2013 Contribution of higher order plasmonic modes on optical absorption enhancement in amorphous silicon thin films *Opt. Commun.* **300** 77–84
[9] Atwater H A and Polman A 2010 Plasmonics for improved photovoltaic devices *Nat. Mater.* **9** 205–13
[10] Amendola V 2016 Surface plasmon resonance of silver and gold nanoparticles in the proximity of graphene studied using the discrete dipole approximation method *Phys. Chem. Chem. Phys.* **18** 2230–41
[11] Mahmoud M A 2014 Optical properties of gold nanorattles: evidences for free movement of the inside solid nanosphere *J. Phys. Chem. C* **118** 10321–8
[12] Mahmoud M A 2014 Plasmon resonance hybridization of gold nanospheres and palladium nanoshells combined in a rattle structure *J. Phys. Chem. Lett.* **5** 2594–600
[13] Lee J, Park J C and Song H 2008 A Nanoreactor Framework of a Au@SiO2 Yolk/Shell Structure for Catalytic Reduction ofp-Nitrophenol *Adv. Mater.* **20** 1523–1528
[14] Hah H J, Um J I, Han S H and Koo S M 2004 New synthetic route for preparing rattle-type silica particles with metal cores *Chem. Commun.* 1012
[15] Priebe M and Fromm K M 2014 One-Pot Synthesis and Catalytic Properties of Encapsulated Silver Nanoparticles in Silica Nanocontainers *Part. Part. Syst. Charact.* **31** 645–651
[16] He Q, Sun H, Shang Y, Tang Y, She P, Zeng S, Xu K, Lu G, Liang S, Yin S and others 2018 Au@ TiO2 yolk-shell nanostructures for enhanced performance in both photoelectric and photocatalytic solar conversion *Appl. Surf. Sci.* **441** 458–65
[17] Li A, Zhu W, Li C, Wang T and Gong J 2019 Rational design of yolk-shell nanostructures for photocatalysis *Chem. Soc. Rev.*
[18] Chen H M, Liu R-S, Asakura K, Lee J-F, Jang L-Y and Hu S-F 2006 Fabrication of nanorattles with passive shell *J. Phys. Chem. B* **110** 19162–7
[19] Anon Reference Air Mass 1.5 Spectra | Grid Modernization | NREL
[20] Chau Y-F C, Lim C M, Chiang C-Y, Voo N Y, Idris N S M and Chai S U 2016 Tunable silver-shell dielectric core nano-beads array for thin-film solar cell application *J. Nanoparticle Res.* **18** 88
[21] Yuan Z, Li X, Guo Y and Huang J 2015 Enhanced absorption of Ag diamond-type nanoantenna arrays *Optoelectron. Lett.* **11** 13–7

AUTHOR INDEX

Institute of Physics
Dirac House, Temple Back
Bristol BS1 6BE UK

ISSN: 1757-8981
ISBN 978-1-7138-3170-9

CURRAN ASSOCIATES INC.
proceedings.com